Nature Aesthetics: Interdisciplinary and Intercultural Dialogue

Series Editor

Shan Gao, Soochow University, Suzhou, China

This book series is devoted to exploring how nature is interpreted and appreciated in different philosophical and cultural traditions around the world, and in turn to expanding the conceptual and practical framework in the field of nature aesthetics. Due to the rapid expansion of urban space and cyberspace alike, we are confronted with a global environmental crisis. Although many interdisciplinary research programs have been established within academic institutions and societies to address the crisis, comparatively little research has been conducted in the field of nature aesthetics, which is now playing an increasingly important role in reshaping the relationship between human beings and nature. This book series is intended to establish a new research platform for promoting an interdisciplinary and intercultural dialogue between East and West in the field of nature aesthetics. It will also address the interrelationship between the theoretical dimension of nature aesthetics and practical aspects, such as rewilding practices, ecological restoration, biocultural conservation, and landscape design.

The series will cover a range of broader topics related to nature aesthetics. These include but are not limited to:

1. Conceptions of Nature, Aesthetics and Ethics in Philosophical and Religious Traditions of East and West
2. Nature Aesthetics and Environmental Humanities
3. Nature Aesthetics and Everyday Life
4. Wilderness Aesthetics and Agricultural Aesthetics
5. Nature Aesthetics, Value, Policy and Law
6. Nature Aesthetics and Environmentalism
7. Nature Aesthetics and Ecotourism
8. Nature Aesthetics and Outdoor Sports
9. Nature Aesthetics and National Park System around the World
10. Nature Aesthetics and Education
11. Nature Aesthetics and Human-Nature Co-Flourishing
12. Comparative Nature Aesthetics
13. Nature Aesthetics and Environmental/Land Art
14. Nature Aesthetics, Politics and Social Theory
15. Nature Aesthetics, Meterverse and Environmental Education
16. Nature Aesthetics and Indigenous Culture

Shan Gao

Comparative Studies Between Emerson and Chinese Philosophy

Nature, Spirituality and Place

Springer

Shan Gao
Department of Philosophy
Soochow University
Suzhou, China

ISSN 3059-4081 ISSN 3059-409X (electronic)
Nature Aesthetics: Interdisciplinary and Intercultural Dialogue
ISBN 978-981-95-0420-6 ISBN 978-981-95-0421-3 (eBook)
https://doi.org/10.1007/978-981-95-0421-3

This Springer imprint is published by the registered company Springer Nature Singapore Pte Ltd.
The registered company address is: 152 Beach Road, #21-01/04 Gateway East, Singapore 189721, Singapore

Preface

China's rapid urbanization and technological advancement are dramatically reshaping the places where people live. With the large-scale migration of populations to urban areas, rural regions experience significant depopulation, facilitating the rewilding of landscapes through natural regeneration, while virtual spaces simultaneously gain increasing prominence. These spatial transformations have tremendous influence on spirituality of Chinese people. For over thousands of years, Chinese philosophy especially Confucianism and Daoism which are generated on agrarian civilization, have greatly shaped the self-identity and spirituality of Chinese culture. The core spirit of Confucianism and Daoism is to live in harmony with nature. Yet, this harmony is fracturing at an alarming pace due to climate change, industrialization, and urban expansion. The moral frameworks of Confucianism and Daoism, once grounded in the spatial and temporal rhythms of agrarian society, have lost their footing, a process exacerbated by the Cultural Revolution's assault on tradition. Confronted with this spiritual crisis, scholars such as Zongsan Mou have sought to revive traditional Chinese culture through Neo-Confucianism,[1] while others, like Xianglong Zhang, propose Confucian cultural reserves.[2] At the governmental level, universities are establishing institutes for traditional culture studies, and rural revitalization initiatives are underway, albeit progressing slowly.

This book does not aim at reconstruction of Confucian or Daoist philosophy to save China's traditional culture. Instead, I view this crisis as an opportunity to reevaluate traditional values and establish new values. Unlike Nietzsche, who in his book titled *the Will to Power* sought to "shatter all past values" through the philosophical concept of will to power,[3] I propose a deconstruction of traditional moral self and values based on deep philosophical analysis of core philosophical categories which shape self-identity and values and reconstruction of the self. In order to achieve

[1] Zongsan Mou, *The Substance of Mind and the Substance of Nature* (Cheng Chung Book Co., 1983).

[2] Xianglong Zhang, "Give Chinese Ancient Endangered Culture a Refuge: Proposal to Establish Confucian Cultural Protection Zones", *Modern Education News,* July 20, 2001, B1.

[3] Friedrich Nietzsche, *The Will to Power,* ed. Walter Kaufmann (New York: Vintage Books, 1968).

the reconstruction, it is necessary to deconstruct the traditional self. This deconstruction entails analyzing the categories that constitute the moral self, tracing their origins, and integrating new dimensions into a redefined identity: the aesthetic self. This aesthetic self rationally apprehends objective values and embodies ecological emotions, a dimension that has been overlooked in China's agrarian history. Ecological aesthetic self and value reshaped the moral virtues which have been shaped by Confucianism and Daoism for thousands of years.

In traditional Confucianism and Daoism, virtue is considered as representing the essence of self, a paradigm I challenge by elevating the ecological aesthetic self as the pinnacle of identity. Globalization and virtual spaces have eroded traditional interpersonal communities rooted in family and kinship, necessitating new forms of human connection. The Confucian ethic of ren (benevolence), built on filial piety and reciprocal moral bonds, no longer suits these reconfigured communities. Human dignity, however, lies in self-transcendence. John Nolt argues, Through systemic support, we can facilitate this transcendence, helping individuals pursue new values over time. My aim is to forge this new self through the cognition, aesthetics, and protection of our lived natural environment, blending theory with practice—including creating more wilderness areas advocating for nature reserves—since theoretical reflection alone lacks vitality.

This book's approach unfolds across three key dimensions. (1) Reexamining spirituality and life's meaning: The quest for life's meaning, intrinsically tied to spirituality, hinges on three core categories: self-transcendence, ultimate value, and power. Viewing Emerson's transcendentalism and Chinese philosophy through this lens reveals their distinct constructions of self-identity, values, and meaning. This ethos, diverging from Christianity's sin-and-salvation narrative, inspired Thoreau and fueled the American Wilderness Protection Act. In contrast, Chinese philosophy constructs meaning within social traditions or nature's spontaneity, fostering the unity of heaven and humanity. These pursuits, various of institutions and cultural place, continue to shape cultural values, influencing aesthetics from American wilderness to Chinese gardens. (2) Exploring the metaphysical interpretations of the concept of nature which have shaped ethical and aesthetic attitudes towards various types of landscapes including wilderness. Roderick Nash asserts, "Civilization created wilderness."[4] In America, transcendentalism underpins a robust wilderness culture, evident in national parks and various wilderness law. China's nascent wilderness culture, spurred by national park initiatives, lacks philosophical grounding in its traditional systems. By comparing Emerson and Chinese philosophy, I explore whether their value systems encompass wilderness, offering insights for new theoretical frameworks without discarding past values as Nietzsche might. (3) Fostering the cultural dialogue between Emerson and Chinese philosophy in their different interpretation of self, virtue, and values. This dialogue can enrich aesthetic values,

[4] Roderick Nash, *Wilderness and the American Mind*, Third edition (Yale University Press, 1982), xiii.

fostering a new natural aesthetic self, distinct from Naess' moral ecological self[5]—that sees nature as an ultimate value, not merely a backdrop for human relations. This self, embracing urban, rural, and wild aesthetics, redefines virtue beyond Confucian interpersonal ethics.

This book is divided into four parts. In Part I, I will introduce the core concepts of this book, namely nature, spirituality, and place, and explain why I use the comparative approach to examine these concepts. The basic logic among the three concepts is that both Emerson and Chinese philosophy regard nature as the bearers of ultimate metaphysical value, which grounds their spirituality. Their pursuit of spirituality deeply shapes their experience of place. I will take the approach of spirituality to conduct the comparative studies between Emerson and Chinese philosophy. In terms of Chinese philosophy, I mainly focus on philosophers of Confucianism and Daoism in the pre-Qin period, especially Confucius, Mencius, and Zhuangzi. Both Ralph Waldo Emerson and these Chinese philosophers lived during periods of profound social transformation. In Emerson's lifetime, traditional religious values such as Calvinism and Unitarianism have gradually lost authority over people. Intellectuals such as Emerson, Margaret Fuller, and Bronson Alcott sought to shape new values through initiatives like utopian communities, advocacy for gender equality, and abolitionist movements. Emerson's primary contribution was his reinterpretation of virtue. He integrates the self with virtue, which responds to the era's meaning crisis by reconstructing a new religion, which is termed as "the romantic triad"[6] by Samantha C. Harvey. It refers to the new trinity among nature, humanity, and spirit. Notably, Emerson draws on Confucianism and Daoism ideas about virtue to address the crisis of meaning in his time. Confucius, Mencius, and Zhuangzi lived during China's Warring States period, a time marked by the disintegration of the Zhou feudal order. Zhou-era values had lost their authority, prompting intellectuals such as Confucius, Mencius, and Zhuangzi to propose diverse solutions to restore social order and meaning in life. This era is well known as the "Hundred Schools of Thought." Confucianism's concept of virtue such as humanity and righteousness and Zhuangzi's virtue of "wandering" emerge as a compelling response to the meaning crisis, offering a framework for reconstructing life's purpose. Today, as artificial intelligence (AI) increasingly shapes daily life, a renewed crisis of meaning has emerged. Through an intercultural dialogue between Emerson and Chinese philosophers, this book explores how new virtues can be formed to address this crisis of meaning.

In Part II and Part III of this book, I will use the core concepts in philosophical study of spirituality to analyze the concept of virtue and its relationship with value of nature. The basic logic is as follows: Emerson and Chinese Philosophy both use virtue to represent the idea of self. Virtue is in essence the transcendence of false self toward the ultimate values of truth, goodness, and beauty. Among these three values, goodness is most important. The values of truth and beauty are based on the goodness. Nature is the bearer of these ultimate values.

---

[5] Arne Naess, The Shallow and the Deep, Long-Range Ecology Movement, *Inquiry*, 1973.

[6] Samantha C. Harvey, *Transatlantic Transcendentalism: Coleridge, Emerson, and Nature* (Edinburgh University Press, 2013).

In Part IV, I will examine the concept of place. In terms of the relationship with nature, spirituality, and place, Emerson and Chinese philosophy's spiritual pursuit shapes their understanding of nature, which has direct influence on their experience of material world. Material world becomes a meaningful place due to their spiritual pursuit of meaning in life.

Suzhou, China Shan Gao

Acknowledgments The writing of this book would not have been possible without the wilderness experiences and philosophical reflections. Professor Ricardo Rozzi's field course titled Tracing Darwin's Path in Chile provided a great opportunity for me to experience the vast wilderness. My philosophical and aesthetic reflections start from that course. This course is part of environmental philosophy and biocultural conservation program coordinated by the University of North Texas, Universidad de Magallanes, and Omora Ethnobotanical Park in the Cape Horn Biosphere Reserve, Chile. That was my first time to encounter with pristine wilderness. This wilderness experience starts my long-term philosophical reflection on the interrelationship among nature, spirituality, and place. Since then, I started to think about topics in natural aesthetics. During my PhD studies in Texas, USA, I organized many wilderness hiking activities in various types of wilderness areas including nature preserve, wild life refuge area, natural area, and state parks, and my friends' participation in those activities also helps me a lot in thinking deeply about wilderness, aesthetic appreciation of nature, and human-wilderness relationships. I would like to thank my friends for their participation in those wilderness trips to many wilderness reserves in Texas: Zongpei Tang, Lei Zhang, Wei Zhang, Wen Hu, and so on.

I am grateful to many professional institutions, academic societies and online lectures for their inspiration for this book. Thanks to the Emerson Society, who provided me the opportunity to present my research on Emerson and Chinese philosophy in the 84th annual gathering of Thoreau Society; I extend my special thinks to Professor Masahiro Morioka, who organized 2nd International Conference on Philosophy and Meaning in Life in Waseda University in Japan in October 7-9. This conference inspires me to think about the concept of spirituality. I also appreciate the online lectures on Emerson which were given by Professor Joseph Urbas, Professor Nicholas Guardiano and Professor Samantha C. Harvey. Thanks to Dr. Li Xu for organizing some of the online lectures on Emerson. I would like to express my thanks to my colleague Professor Huanzhong Han whose research on the integration of Confucianism, Daoism, and Buddhism, as well as many academic activities he organized in many famous Buddhism temples in Suzhou such as Hanshan Temple and Chongyuan Temple provide great inspiration in the comparative study of Emerson and Chinese philosophy in this book. I would also like to express my deepest appreciation for my husband's entire family, especially my mother-in-law, brother-in-law and sister-in law, and my family for their long time support for my life so that I can spend more time in philosophical thinking.

Finally, thanks to the cover provider for this book, Hua Song.

Competing Interests The author has no competing interests to declare that are relevant to the content of this manuscript.

Contents

Part I
Introduction

In this part, I will introduce the philosophical framework of this book. It is divided into four chapters which will focus on four questions: (1) Why start from the concept of nature? I argue that the value dimension of nature is the primary focus on this book and it is the foundation for people's spirituality and nature conservation; (2) What is spirituality? In this part, I will focus on four dimensions of spirituality including pursuit of good life, value, self-transcendence and spiritual practices. (3) Why choose Emerson to conduct comparative studies with Chinese philosophy? I will introduce Emerson as the founder of American transcendentalism and its close relationship with the theme of the book. (4) Why making comparative studies between Emerson and Chinese philosophy? I will discuss three foundations for conducting the comparative studies between Emerson and Chinese philosophy, especially Confucianism and Daoism.

Chapter 1
Nature

Nature, spirituality, and place are three interrelated key concepts I would like to focus on when conducting comparative studies between Emerson and Chinese philosophy. Among the three concepts, the basic logical relationship is as follows: Spirituality in essence is the pursuit of meaning in life through self-transcendence towards ultimate values. Many philosophers interpret nature as the bearers of ultimate values including truth, goodness, and beauty. For this reason, their pursuit of spirituality is deeply connected to nature. Spirituality has direct influence on sense of place. Various forms of land such as farming land, gardens, and wilderness becomes sacred in different cultures. In this chapter, I will first introduce why I start from the concept of nature and then give a review of Western philosophical history in their interpretation of nature and its relationship with value. With the historical background in Western philosophers' interpretation of nature, we can gain a deep understanding how Emerson's new interpretation of nature and its relationship with value has shaped the spirituality in American transcendentalism and how it generates the sense of place which has been neglected by many Western philosophers for long time.

The reason why I started from nature is due to the practice of creating national parks in China. In 2015, China proposed to establish the national park system pilot starting from nine provinces. In 2020, China began to end the pilot and officially established national parks. However, within the areas protected by national parks, wilderness has not yet become a highly regarded place. Wilderness is not a legal concept in China, and it lacks cultural elements, so there is no mention of wilderness protection areas. Professor Rui Yang and Yue Cao from Tsinghua University published an article in the Guangming Daily, "Promoting China's Wilderness Protection with the Opportunity of National Parks." The article points out that China has one of the largest areas of existing wilderness in the world, but there is no strict wilderness protection. They suggest designating "wilderness protection zones" in the functional zoning of national parks, and implementing "the strictest protection" in the form of wilderness for the most core essence of national parks. However, both wilderness and national parks are concepts mainly originated from America.

S. Gao, *Comparative Studies Between Emerson and Chinese Philosophy*, Nature Aesthetics: Interdisciplinary and Intercultural Dialogue,
https://doi.org/10.1007/978-981-95-0421-3_1

Although China has brought the American best idea of national parks into practice, it lacks the support of Chinese cultural values, which are deeply influenced by Confucianism and Daoism. Without this support, vast wilderness areas in China will not be effectively protected.

Protecting the wilderness requires a new philosophy of wilderness. However, the dominating Chinese spirit of living in harmony with nature tend to neglect the idea of wilderness philosophy. It needs new development to generate such kind of wilderness philosophy in China. However, this book does not undertake this grand project, that is, to reconstruct China's theory of harmony between man and nature, to lead it towards the new theory of living in harmony with nature, in which wilderness is included and valued. Instead I will start from comparative studies between Emerson's transcendentalist philosophy that directly influenced American wilderness values and wilderness protection and Chinese philosophy. I will explore how their spirituality is deeply related to the interpretation of nature and how this spirituality has tremendous influence on their sense of place. In Emerson, wilderness is valued, while Chinese philosophy promotes attachment to the land. This is the first step in constructing the new theory of living in harmony with nature within the background of Chinese philosophical tradition. I will take the approach of intercultural dialogue to do it. If the thought of harmony between man and nature that has influenced China for thousands of years cannot get new development, then the protection of China's vast wilderness areas cannot be fundamentally guaranteed. Although many philosophers in America has produced theory of wilderness philosophy such as Holmes Rolston III, their theory is founded on the framework of Western philosophy and religion, some basic philosophical concepts cannot be applicable to China. We need to start from China's traditional culture and develop new thoughts on this basis. However, the cultural dialogue is very necessary for the self-reflection of our own cultural understanding of nature and developing it based on the reflection.

Lynn White, a historian of medieval American history, proposed a similar view in his famous article "The Historical Roots of Our Ecological Crisis" published in the journal of *Science*. He believes that the anthropocentric attitude originates from the following two historical developments: (1) The Western modern democratic system originating from conquest, plunder, and colonial development. (2) The Judeo-Christian tradition, which believes that all created things are for human purposes, and God chooses humans to rule everything on earth. Lynn White believes that the second type of religious view constitutes the root of the ecological crisis brought about by anthropocentrism. He believes that since religion is the root of ecological crisis, our solution should also be religious, such as Saint Francis of Assisi, a radical Christian, who provides us with a non-anthropocentric religion of equality for all things. The harmony between man and nature in Chinese philosophy is not religious but a philosophical tradition with spirituality. To some extent, this philosophical tradition cannot solve China's wilderness protection problem. To solve this problem, new development is needed on this basis. And the key to development is to solve some fundamental philosophical questions: Does the concept of nature in Chinese culture accommodate wilderness? How to define wilderness? What is value? What is the relationship between wilderness and value? Does the value in wilderness justifies our moral obligation to it? To answer these questions, interdisciplinary knowledge is required. However, in this chapter, I only focus on the value dimension of nature and wilderness and how this dimension is deeply related to spirituality of human beings and their sense of place. I will start from tracing the concept of nature and relationship with value in the history of Western philosophy, and it is the foundation for understanding how Emerson's idea of nature has developed from the historical background and initiate an aesthetic revolution and create a new sense of place.

In the history of Chinese and Western philosophy, human interpretation of nature is divided into three types. The first type is to combine nature with value, believing that nature itself has value. Philosophers representing this type include Aristotle, Rousseau, and Emerson. The second type of interpretation of nature by philosophers is to separate nature from value, and philosophers representing this type of interpretation include Descartes, Spinoza, and Kant. The third type is to see nature as a product of interpretation, and philosophers of this type include Nietzsche.

In terms of the relationship between nature and value, there are three types of interpretation in the history of Western philosophy. (1) Combination of Nature and Value. Nature is the combination of ultimate value, which is truth, goodness, and beauty; (2) Separation of Nature and Value. Nature is not the bearer of ultimate value; rather ultimate value lies in human beings; (3) Attribution value to Nature. Although nature doesn't have value. However, we can attribute value to nature so that we can find the meaning in life. Based on the examination, we will find there is a trend in the history of Western philosophy that philosophers tend to ignore the importance of material nature and sense of place in their spirituality. Even though some philosophers' interpretation of nature is connected to values, their metaphysics cannot support valuing physical nature and a sense of place. In the following, I will examine the metaphysical concept of nature and its relationship with value. At the same time, I will discuss why the metaphysical concept of nature has caused the neglect of the philosophical importance of material nature and place.

1 The Combination of Nature and Value

Many Western philosophers' interpretations of nature are related to values. These philosophers include Aristotle, Plotinus, Augustine, and Rousseau. By values, I refer to ultimate values of truth, goodness and beauty in the metaphysical sense. Different philosophers stress different aspects of these values.

1.1 Aristotle: Nature and Goodness

Aristotle's concept of nature is closely related to the value of goodness. By "nature", he refers to two things: matter and form. For him, form is more important than matter. It can represent nature.

He points out that "form is a more plausible candidate for being nature than matter is because we speak of a thing as what it actually is at the time, rather than what it then is potentially."[1] In terms of the concept of form, it can be best understood through Aristotle's four causes of nature. The four causes are the material cause, the formal cause, the final cause, and the efficient cause. The last three causes correspond to the concept of form. There are three dimensions in the meaning of the form: (1) It refers to the principle that defines a thing's identity. It is also the formal cause of nature; (2) It is related to an original source of motion and change, which is the efficient cause of nature; (3) It is related to the purpose. The purpose is the final cause of nature. I will focus on the final and efficient cause of nature since they are closely related to the value of goodness. In terms of the final cause, he writes, "Starting from a given source of change does not result in the same end in every case, but it is not just any chance end either; there is in fact always a tendency towards the same end, unless something intervenes."[2] This teleological tendency underscores Aristotle's belief that nature operates purposefully.

In his biological works, particularly *Parts of Animals*, he reiterates this principle: "Now, as each of the parts of the body, like every other instrument, is for the sake of some purpose, viz. some action, it is evident that the body as a whole must exist for the sake of some complex action."[3] For instance, he observes serrated-toothed animals and remarks that, "No animal has saw-teeth as well as tusks; for Nature never does anything without purpose or makes anything superfluously."[4] Similarly, in analyzing fish, he states, "Fishes have no separate limbs attached to the body. (a) This is because Nature never makes anything that is superfluous or needless, and by their essence and constitution fishes are naturally swimmers and so need no such limbs."[5] For Aristotle, the purpose is the end and also the final cause of nature. It has the value of goodness.

Aristotle's concept of the efficient cause, the agent of motion, also converges with the form. Aristotle explains, "The agent of change will aways bring with it some form, which will either be 'such-and-such a thing' or 'such-and-such a quality' or 'such-and-such a quantity', and this form will be the principle and cause of any change that the agent of change produces. For instance, it is an actual man who creates a man out of that which is potentially a man."[6] Here, the form as the efficient cause initiates movement toward its realization, aligning the efficient cause with the final cause's purpose. For the efficient cause, Aristotle employs the term of "the unmoved mover" to express the source of all motion and change in the cosmos. This eternal, unchanging entity drives natural processes without itself being moved, unifying the teleological structure of nature with a metaphysical foundation. The relationship between nature and value emerges here: the formal cause, unified with purpose and

---

[1] Aristotle, *Physics*, trans. Robin Waterfield (Oxford University Press, 1996), p. 35.

[2] Ibid., p 53.

[3] Aristotle, *Parts of Animals*, trans. A. L Peck (Harvard University Press, 1961), p. 103.

[4] Ibid., p. 211.

[5] Ibid., pp. 417–418.

[6] Aristotle, *Physics*, trans. Robin Waterfield (Oxford University Press, 1996), p. 60.

motion, orients natural objects toward their good, while the unmoved mover anchors this goodness in a cosmic order.

Aristotle's concept of the Unmoved Mover is the efficient cause of all things and it has a close relationship with the value of goodness. This Unmoved Mover is also termed as God. He writes, "And God *is* in a better state. And life also belongs to God; for the actuality of thought is life, and God is that actuality; and God's essential actuality is life most good and eternal. We say therefore that God is a living being, eternal, most good, so that life and duration continuous and eternal belong to God; for this *is* God."[7] In Aristotle's view, God is not a personified God. God does not reward good and punish evil, nor is he jealous. God is something with eternal life, but this life is not the organic life but a life of thought. It is an origin of all things, and the origin is most perfect and intact. God moves all objects in nature through his thoughts. Thoughts are ultimate good and it is the metaphysical foundation for the goodness of form in nature.

From the above explanation, it can be seen that Aristotle's concept of nature is not a collection of facts. Instead, it is in essence a form, which is deeply interconnected to value of goodness. This is an important reason why Aristotle regards the contemplative life as the best way of life. It also explains why Aristotle neglects the concept of place in his philosophy although his concept of nature is related to the value of goodness. For him the material aspect of nature is considered as less important than the other three causes of nature. His best way of life is not related to attachment to material places rather to thoughts.

1.2 Plotinus: Nature, Goodness, and Beauty

Plotinus' concept of nature is deeply connected to values of goodness and beauty. For him, all things in nature have their own nature and conditions and they serve the life of the whole, and also have the ability to receive. He writes, "Everything in the universe serves the universe according to its own nature and characteristics, affects the universe and is affected by the universe, just like the various limbs and organs of individual living things, according to their own nature and conditions, assist the life of the whole, serve the purpose of the whole life, and at the same time have their own special duties and functions. Each part gives out all it has, while receiving all from other parts, because they all have a certain ability to receive. There is a common consciousness between them, which connects them."[8] It can be seen that in Plotinus, the main reason why all things in nature have value is that the ultimate entity that produces all things in nature is the ultimate good, so every produced object shares this good, which brings value to nature. However, it needs to be emphasized that this value is not originally possessed by the material nature, because the material

---

[7] Aristotle, *Metaphysica* in *The Works of Aristotle*, vol. VIII, trans. W. D. Ross (Oxford at the Clarendon Press, 1928), 1072 b.

[8] Plotinus, *The Ennead*, trans. Stephen MacKenna (Penguin Books, 1991), p. 46.

represents evil in his view. However, because the material has form and soul from reason and thus has life, this life covers evil.

First, for Plotinus, nature is beautiful. The beauty of nature lies not in the characteristics of natural objects. Instead, the cause of the beauty in nature is attributed to the soul, which brings life to natural objects. He criticizes some popular views that the reason why some objects are pleasing to the eye is because of the appropriate proportion between the parts of the object and the whole, and if it is a single thing, such as beautiful colors and light, it is not beautiful. Plotinus criticizes this view with an example, "Again since the one face, constant in symmetry, appears sometimes fair and sometimes not, can we doubt that beauty is something more than symmetry, that symmetry itself owes its beauty to a remoter principle?"[9] He then continues to ask, where exactly is the proportion of beauty in the outstanding knowledge or virtues of law or various disciplines? He uses these examples to show that beauty is not a proportion, but is related to the soul. "Our interpretation is that the Soul – by the very truth of its nature, by its affiliation to the noblest Existens in the hierarchy of Being – when it sees anything of that kin, or any trace of that kinship, thrills with an immediate delight, takes its own to itself, and thus stirs anew to the sense of its nature and of all its affinity."[10] The quotation indicates that the human capacity to appreciate the beauty of nature arises from a fundamental correspondence between the human spirit and the soul of nature, both of which participate in the same universal spiritual principle and therefore exhibit essential affinities. In this framework, beauty does not reside in raw material but the emergent quality of the form. Material objects devoid of the shaping forces of form appear disordered or aesthetically unpleasing. Beauty attaches itself to material existence only when the latter attains a state of harmonious unity. Thc formation of beautiful objects is due to the sharing of some constitutive power, but the power of constitution comes from the divine form.

Second, Plotinus thinks that nature is good. He says, "Now, no one of any intelligence complains of these others, man's inferiors, which serve to the adornment of the world;"[11] For him, other lives in nature all have their own beautiful purposes, they are indispensable existences on the earth, so people should not blame the cruelty of animals. This beautiful purpose is good, and the ultimate source of this good is the One. Both humans and non-human existences in nature share the One. Plotinus emphasizes that if other living things live a life in accordance with their nature, then they also possess the same good as humans. He states, "For example, animals with musical talent, if they are in favorable circumstances and can sing in a way that is in accordance with their nature, then they are living the life they want. Similarly, we think that welfare is a kind of purpose, which is the final stage of natural desire."[12] For Plotinus, as long as nature runs through other living things, they have realized their own good. Even many behaviors in the animal kingdom that are very cruel in human eyes, such as predation and being preyed upon, are also good, because this

[9] Plotinus, *The Enneads*, trans. Stephen MakKenna (Penguin Books, 1991), p. 46

[10] Ibid., p. 47.

[11] Ibid., p. 146.

[12] Ibid., p. 55.

is exactly the life that conforms to the nature of animals. For him, both animals and plants realize a form of life that accords with their inherent nature. This principle extends even to the simplest living organisms, which can possess a good of their own. Their existence may be vigorous and fruitful or, alternatively, withered and barren. Yet a life that conforms to its inherent nature remains good in itself. Plants, in particular, cannot be denied such a good simply on the grounds that they lack sentience or feelings. In this sense, to have a good of one's own is to possess welfare.

1.3 Augustine: Nature and Beauty

According to Augustine, nature is beautiful. But the beauty of nature is just an imperfect manifestation of the beauty of God, and can not be compared with God. This view is influenced by Plotinus. Plotinus writes, "So, mounting, the Soul will come first to the Intellectual-Principle and survey all the beautiful Ideas in the Supreme and will avow that this is Beauty, that the Ideas are Beauty. For by their efficacy comes all Beauty else, by the offspring and essence of the Intellectual-Being."[13] Augustine sees worldly beauty as imperfect and transient. He argues that while nature's allure reflects God's creation, it decays, urging one to seek beyond it, "Late have I loved you, beauty so old and so new: late have I loved you. And see, you were within and I was in the external world and sought you there, and in my unlovely state I plunged into those lovely created things which you made. You were with me, and I was not with you. The lovely things kept me far from you, though if they did not have their existence in you, they had no existence at all. You called and cried out loud and shattered my deafness. You were radiant and resplendent, you put to flight my blindness. You were fragrant, and I drew in my breath and now pant after you. I tasted you, and I feel but hunger and thirst for you. You touched me, and I am set on fire to attain the peace which is yours."[14] For Augustine, nature's beauty reveals God's creation, mediated by the Holy Spirit, forging an intimate bond between humanity and the divine. However, many people tend to love God's creation and forget about the creator. For Augustine, it is morally wrong. In the following, I will discuss Augustine's interpretation of nature's beauty and its relationship with God from three dimensions: (1) order in nature's beauty; (2) nature as a medium to know God; (3) contemplation of God for aesthetic appreciation of nature.

First, Augustine views nature's beauty as an ordered hierarchy within God's creation, encompassing inanimate objects, seasons, and living beings. He writes, "For things earthly were neither to be made equal to things heavenly, nor were they, though inferior, to be quite omitted from the universe. Since, then, in those situations where such things are appropriate, some perish to make way for others that are born in their room, and the less succumb to the greater, and the things that are overcome are transformed into the quality of those that have the mastery, this is the appointed order

[13] Plotinus, *The Ennead*, trans. Stephen MacKenna (Penguin Books, 1991), p. 55.

[14] Augustine, *Confessions*, trans. Henry Chadwick (Oxford University Press, 1991), p. 201

of things transitory."[15] For Augustine, these lesser beauties—transient and subordinate—fit within God's cosmic design. Yet human habits obscure this whole, skewing perception toward comfort, "Therefore it is not with respect to our convenience or discomfort, but with respect to their own nature, that the creatures are glorifying to their Artificer."[16] Each entity, with its genus and species, boasts internal harmony, praising God irrespective of human enjoyment. Contemplating this order exalts the Creator.

Second, nature is seen as a medium to know God. The nature as understood by Augustine doesn't have intrinsic value as the environmental philosophers have argued. Instead, nature is treated as the medium for human beings to know God. The following passage from his *Confessions* can better illustrate this point, "'Tell me of my God who you are not, tell me something about him.' And with a great voice they cried out: 'He made us'. My question was the attention I gave to them and their response was their beauty."[17] This beauty, ever-changing, points to eternal beauty in God: "'O God of hosts, turn us and show us your face, and we shall be safe'. For wherever the human soul turns itself, other than you, it is fixed in sorrows, even if it is fixed upon beautiful things external to you and external to itself, which would nevertheless be nothing if they did not have their being from you."[18] For Augustine, nature doesn't have its own intrinsic value or goodness and its existence depends on God. Although nature is beautiful, its beauty is inferior to God's beauty which is based on goodness.

Third, for Augustine, contemplating God is metaphysically important, which can help people appreciate nature correctly. For him, it is the wrong order if we appreciate the beauty of nature first. The reason can be found in this passage, "To this I may add another form of temptation, manifold in its dangers. Beside the lust of the flesh which inheres in the delight given by all pleasures of the senses (those who are enslaved to it perish by putting themselves far from you), there exists in the soul, through the medium of the same bodily senses, a cupidity which does not take delight in carnal pleasure but in perceptions acquired through the flesh. It is a vain inquisitiveness dignified with the title of knowledge and science."[19] For Augustine, God speaks to the soul's nobler part not through material forms but through truth. He stresses that we should contemplate nature through God first rather than appreciate the beauty of nature through senses. For him, the sensual appreciation of nature will easily make people addicted to the pleasure of the senses. In this way, we tend to use nature for the human beings' welfare. He writes, "Thus the reason of one contemplating nature prompts very different judgments from those dictated by the necessity of needy, or the desire of the voluptuous; for the former considers what value a thing in itself has in the scale of creation, while necessity considers how it meets its need; reason looks for what the mental light will judge to be true, while pleasure looks

[15] Augustine, *The City of God*, vol. 1, ed. Marcus Dods (T&T Clark, 1913), p. 485.

[16] Ibid., 486.

[17] Augustine, *Confessions*, trans. Henry Chadwick (Oxford University Press, 1991), p. 183.

[18] Ibid., p. 61.

[19] Ibid., pp. 210–211.

for what pleasantly titillates the bodily sense."[20] From this passage, we can find that Augustine stresses contemplation of nature based on the mental light, which is from God. Without the guidance from God, contemplation of nature may lead us towards wrong direction. From the above, we can see that Augustine doesn't have attachment to place since his real place or home lies in God. Material nature for him doesn't have value for their own sake. Rather it only has second value which is derived from God.

1.4 Rousseau: Nature, Good, and Beauty

Jean-Jacques Rousseau regards nature as inherently valuable, attributing this worth to God as its source. He infers God's existence from the motion of matter, a view articulated through the Savoyard Vicar in *Emile*. The Vicar asserts that the movement of inanimate objects cannot originate within matter itself but stems from a spontaneous will: "Inanimate bodies act only by motion, and there is no true action without will."[21] This will operates not arbitrarily but according to laws, suggesting a guiding wisdom: "If moved matter shows me a will, matter moved according to certain laws shows me an intelligence."[22] For Rousseau, God is not merely the architect of order but imbues it with value—harmony, beauty, and a teleological thrust toward ultimate goodness. He contrasts this divine coherence with human disarray: "The picture of nature had presented me with only harmony and proportion; that of mankind presents me with only confusion and discorder! Concert reigns among the elements, and men are in chaos! The animals are happy; their king alone is miserable!"[23] This judgment emerges not from abstract reasoning but from observing nature's unity, exemplified by the Vicar's watch analogy. He reflects, "...all these wheels are moving in harmony for a common end which it is impossible for me to perceive."[24] recognizing purpose in design despite not grasping its full intent. Rousseau contends that this order bears moral and aesthetic dimensions, reflecting a merciful God's love.

Rousseau further elaborates that God's wisdom, power, and will flow from this love, aiming for goodness: "I judge that there is an order in the world although I do not know its end; to judge that there is this order it suffices for me to compare the parts in themselves, to study their concurrences and their elations, to note their harmony."[25] Here, love assumes metaphysical primacy—God's affection for order precedes and shapes its creation, with goodness as its ultimate end. Nature's value thus elevates its role in Rousseau's vision of the good life. In Emile, he celebrates

[20] Augustine, *The City of God*, vol. 1, ed. Marcus Dods (T&T Clark, 1913), p. 456.

[21] Jean-Jacques Rousseau, *Emile or On Education*, trans. Allan Bloom (New York: Basic Books, Inc., Publishers, 1979), p. 273.

[22] Ibid., p. 275.

[23] Ibid., p. 278.

[24] Ibid., p. 275.

[25] Ibid.

its rejuvenating power: "In spring, the fields are nearly barren, nothing grows, the woods lack shade, the grass barely sprouts, yet our hearts are stirred by this scene. Seeing nature return to the earth, we feel our own life revived; we are enveloped by a scene of joy; our tender companion beholds it, sweet tears of affection welling up to our eyelids; yet, at the grape harvest, so jubilant and lively, we watch without shedding a tear."[26] Likewise, in *Reveries of the Solitary Walker*, Rousseau expresses his sentiment of contentment and peaceful joy when he is alone in pristine nature: "What do you enjoy in such a situation? Nothing external to ourselves, nothing if not ourselves and our own existence. As long as this state lasts, we are sufficient unto ourselves, like God. The sentiment of existence, stripped of any other emotion, is in itself a precious sentiment of contentment and of peace which alone would suffice to make this existence clear and sweet to anyone able to spurn all the sensual and earthly impressions which incessantly come to distract us from it and to trouble its sweetness here-below."[27] Rousseau's aesthetic feeling in being alone with nature is the result of his metaphysical understanding and interpretation of nature. For him, nature represents the ultimate value of goodness and beauty.

2 Separation of Nature and Value

Some philosophers separate value from nature in their interpretation of nature. They hold that nature represents the truth. However, the truth they stress doesn't have value dimension. This interpretation brings three different types of attitudes towards naturc: (1) Knowing nature forms an important part of the good life; Epicurus and Spinoza represent this view; (2) Knowing nature serves to conquer and master it. Descartes exemplifies this position; (3) The aesthetic appreciation of nature can improve morality, although nature itself is not the bearer of truth, goodness, and beauty. Kant represents this perspective.

2.1 Epicurus: Understanding Nature to Eliminate Fear

For Epicurus, nature itself doesn't have value for its own sake. However, we should still try to understand nature's workings so that this knowledge of nature can help us eliminate fear. He emphasizes that the ultimate reality is only objects and void. He criticizes various viewpoints that hold that sacred consciousness exist outside objects and void. For example, some people think that there is an existence that enjoys complete immortal happiness in the movement, return, and rise and fall of the heavens and the earth, and this existence is diligently managing all

[26] Ibid., pp. 167–168.

[27] Jean-Jacques Rousseau, *Reveries of the Solitary Walker*, trans. Peter France (London: Penguin Classics, 1979). p. 89.

this. But Epicurus thinks that happiness is incompatible with worry and favoritism. The fixed and unchanging cyclical movement phenomenon of celestial bodies comes from the initial aggregation and combination of atoms when the world was created. Regarding the overall existence of objects, Epicurus believes that some of them are composites, and some are elements of composites. Elements are indivisible atoms and unchangeable, so composites will not be destroyed into nothing after disintegration. He believes that as the basic elements of all things, atoms are indivisible and physical things. Composites are produced by atoms and return to atoms when they are destroyed. Below, I will explain Epicurus' view of atoms from three aspects and at the same time discuss Epicurus's understanding of the value of nature in this process.

Epicurus characterizes atoms as continuously moving, with modes of motion including vertical motion, deviation from vertical motion, and vibration within compounds. He emphasizes that the number of atoms is infinite and they will go further in motion, so it is possible to have an infinite number of worlds. Although the movement of atoms is different in mode, it is inevitably equal in speed. As long as there are no obstacles, the speed of heavy atoms and small and light atoms is the same. However, the movement of different atomic compounds is different, because the atoms in the compound will collide due to movement in different directions, thus slowing down the overall movement speed. Regarding the characteristics of atomic motion, there is another crucial point, that is, atoms will deviate in motion. Lucretius elaborated on this point when explaining Epicurus' atomic theory. "This point too herein we wish you to apprehend: when bodies are borne downwards sheer through void by their own weights, at quite uncertain times and uncertain spots they push themselves a little from their course: you just and only just can call it a change of inclination. If they were not used to swerve, they would all fall down, like drops of rain, through the deep void, and no clashing would have been begotten nor blow produced among the first-beginnings: thus nature never would have produced aught."[28] Epicurus believes that when atoms fall in the void, although the speed is the same, there will still be deviations, so new objects can be produced.

Regarding the movement and characteristics of atoms, two points are crucial. First, he emphasizes that the motion of atoms lacks any purposive or teleological character. The fundamental constituents of things, collide with one another in various ways, thereby giving rise to the earth, the seas, the heavens, and the living beings that reproduce across generations. This process of collision, however, remains entirely without direction or purpose. Second, atoms do not possess other attributes, such as color and sound, etc. These attributes will change constantly, but the change is only a change in position and quantity, and there will be no change in phenomena. From these two points, it can be seen that Epicurus regards the basics of things as independent of any sensory characteristics, and only exists due to the changes in position and shape differences brought about by mathematical and geometric characteristics. The shape of atoms is diverse, but not infinitely diverse. It can be

[28] *Great Books of the Western World* (12), ed. Robert Maynard Hutchins (EncyclopÆdia Britannica Inc. 1952), p. 17.

seen that when atoms do not possess the attributes of color and sound, and do not have purposefulness, nature composed of atoms itself does not have the ultimate value meaning of truth, goodness, and beauty. But Epicurus emphasizes that people should understand nature. The purpose is to get rid of various fears and to achieve peace of mind. The purpose of studying nature is to serve the purpose of studying human nature. In a letter to Pythocles, he writes, "And most of all give yourself up to the study of the beginnings and of infinity and of the things akin to them, and also of the criteria of truth and of the feelings, and of the purpose for which we reason out these things. For these points when they are thoroughly studied will most easily enable you to understand the causes of the details."[29] He believes that studying nature is to better serve human nature. He emphasizes this point many times in the basic way. He writes, "If we were not troubled by our suspicions of the phenomena of the sky and about death, fearing that it concerns us, and also by our failure to grasp the limits of pains and desires, we should have no need of natural science."[30] "A man cannot dispel his fear about the most important matters if he does not know what is the nature of the universe but suspects the truth of some mythical stroy. So that without natural science it is not possible to attain our pleasures unalloyed."[31]

It can be seen that according to Epicurus, the only purpose of human understanding of nature is to serve human nature and get rid of the fear brought by ignorance. This is essentially different from Emerson. In Emerson, people have an obligation to understand nature, because nature has the ultimate meaning of truth, value, and beauty. The role of natural science is not to get rid of people's fear, but to enhance human beings' virtue.

2.2 Descartes: Understanding Nature to Conquer Nature

When Descartes interprets nature, he focuses on scientific knowledge including maths, geometry and physics. For him, the science can help people know the truth of nature. Descartes' understanding of truth has two levels. The first level of truth is self-evident intuition. It is expressed in his famous philosophical statement, "I think, therefore I am." By "I", he refers to the "thinking thing", a thing that can reason, understand, affirm, deny, and etc. The second level of truth is the one which is given by sciences. For him, the sciences are founded on human reason which is the first truth he never doubts. In interpretation of natural objects, he stresses maths and geometry.

For Descartes, nature can be explained in terms of objective truth, which can be expressed by a few principles. "After this, passing my mind again over all the objects that have ever presented themselves to my senses, I daresay I did not notice

[29] *Epicurus: The Extant Remains*, trans. Cyril Bailey, M.A. (Oxford: At The Clarendon Press, 1926), p. 81.

[30] Ibid., p.97.

[31] Ibid.

anything in them that I could not explain easily enough by means of the principles I had found. But I must also admit that the power of nature is so ample and so vast, and these principles are so simple and so general, that I notice hardly any particular effect without at once knowing that it can be deduced in many different ways from them, and that ordinarily my greatest difficulty is to find in which of these ways it depends on them."[32] In addition to science, Descartes also stresses the importance of experiments in our understanding of the truth in nature. For him, these scientific experiments can help people gain more knowledge of nature: "Moreover, I noticed, in regard to experiments, that they are the more necessary as one is more advanced in knowledge."[33]

One point I would like to stress here is that Descartes doesn't aim at acquiring truth for its own sake. He has explicitly expressed this view in the following passage, "But as soon as I had acquired some general notions regarding physics, and, beginning to test them in various particular difficulties, I had noticed where they could lead and how much they differ from the principles that have been in use up to the present, I believed I could not keep them hidden away without sinning grievously against the law that obliges us to procure, as much as is in our power, the common good of all men. For these notions made me see that it is possible to arrive at knowledge that would be very useful in life and that, in place of that speculative philosophy taught in the schools, it is possible to find a practical philosophy, by means of which, knowing the force and actions of fire, water, air, the stars, the heavens, and all the other bodies that surround us, just as distinctly as we know the various skills of our craftsmen, we might be able, in the same way, to use them for all the purposes for which they are appropriate, and thus render ourselves, as it were, masters and possessors of nature."[34] For Descartes, to know the certainty of nature is for the sake of utility. The final purpose for him is the welfare of human beings. Nature itself doesn't have value for its own sake and it only has instrumental value.

Here, a brief comparative study between Descartes and Aristotle can gain a deeper understanding of both philosophers' views on nature. For Aristotle, the strong sense of wonder will naturally motivate people to know the truth in nature. For him, contemplation of the truth of nature can make people live a best way of life. However, for Descartes, we know the certainty and truth in nature just for the final purpose of improving human beings' welfare. For him, the final value lies in human beings' welfare. The truth of nature as explained by math, geometry and physics doesn't have any value dimension. It is not beautiful or moral. It is just the objective truth and human beings use this truth for benefiting their lives.

---

[32] René Descartes, *Meditations on First Philosophy*, trans. Donald A. Cress (Hackett Publishing, 1998), p. 36.

[33] Ibid.

[34] Ibid., pp. 34–35.

2.3 *Spinoza: Understanding Nature to Achieve the Supreme Good*

In Spinoza's philosophy, nature and value are separate. Nature in Spinoza's view does not possess the ultimate values of truth, goodness, and beauty. He strongly criticizes the use of these value terms to describe nature. He writes, "After men persuaded themselves that everything which happens, happens on their account, they had to judge that what is most important in each thing is what is most useful to them, and to rate as most excellent all those things by which they were most pleased. Hence, they had to form these notions, by which they explained natural things: *good, evil, order, confusion, warm, cold, beauty, ugliness.* And because they think themsleves free, those notions have arisen: *praise* and bla*me*, *sin* and *merit*."[35]

He distinguishes *natura naturans* from *natura naturata.* The former refers to active nature and the latter passive nature. He has given a clear definition for them: (1) active nature: "what is in itself and is conceived through itself, or such attributes of substance as express an eternal and infinite essence."[36] (2) passive nature: "whatever follows from the necessity of God's nature, *or* from any of God's attributes, that is, all the modes of God's attributes insofar as they are considered as things which are in God, and can neither be nor be conceived without God."[37] The term God is alternative expression for *natura naturans.* For better understanding the concept of God, let's take a look at the following quotes about Spinoza's account of God.

> God is not only the cause of things' beginning to exit, but also of their persevering in existing, *or* (to use a Scholastic term) God is the cause of the being of things.[38] Whatever is, is in God, but God cannot be called a contingent thing. For he exists necessarily, not contingently.[39] God's power is his essence itself.[40] God loves himself with an infinite intellectual love.[41] …things have been produced by God in highest perfection, since they have followed necessarily from a given perfect nature.[42] Nature has no end set before it, and that all final causes are nothing but human fictions.[43]

First, rational creativity as the source of all things. Spinoza regards intellect as the origin of creativity, positing that the power to create and sustain all things emanates from God's nature, which he equates with reason. While he offers no explicit definition of reason, Spinoza elucidates it through comparative and relational analyses. He asserts, "So God's intellect, insofar as it is conceived to constitute God's essence, is

---

[35] Benedict De Spinoza, *A Spinoza Reader: The Ethics and Other Works*, trans. Edwin Curley (Princeton University Press 1994), p. 113.

[36] Ibid., pp. 104–105.

[37] Ibid., p. 105.

[38] Ibid., p. 102.

[39] Ibid., p. 104.

[40] Ibid., p. 109.

[41] Ibid., p. 260.

[42] Ibid., p. 107.

[43] Ibid., p. 111.

really the cause both of the essence and of the existence of things."[44] This positions God's reason as ontologically prior to all entities, the foundational cause of their form and truth, in contrast to human reason, which emerges subsequent to creation. Spinoza further aligns God's reason with His will, arguing that creation follows necessarily from God's nature, not arbitrary whim. He rejects popular notions of divine omnipotence as capricious, stating, "God can bring it about that it would not follow from the nature of a triangle that its three angles are equal to two right angles; *or* that from a given cause the effect would not follow-which is absurd."[45] God's reason, will, and power are thus identical "The intellect, power, and will by which God creates, understands, preserves, or loves created things are not distinct from one another, except in our thought."[46] This conception resonates with earlier thinkers, notably Giordano Bruno, who in his book titled *On Cause, Principle, and Unity* frames universal reason as a creative force. Bruno writes, "The universal intellect is the innermost, most real and most proper faculty or potential part of the world soul. It is that one and the same thing that fills everything, illuminates the universe and directs nature to produce her various species suitably. It is to the production of natural things what our intellect is to the production of the representations of things."[47] Bruno's universal intellect, like Spinoza's divine reason, systematically generates nature's diversity, analogous to human conceptual production, yet remains value-neutral, eschewing categories such as good, evil, or beauty. Spinoza's frequent recourse to mathematical and geometric methods in the Ethics reinforces this rational, non-teleological framework.

Second, the purposelessness of divine creativity. Spinoza insists that God's creative act lacks purpose, driven solely by intrinsic necessity. He declares, "Gods acts from the laws of nature alone, and compelled by no one."[48] This self-caused necessity precludes external goals: creation presupposes nothing beyond God Himself. Challenging the teleological biases of his contemporaries, who imbued nature with value-laden purposes, Spinoza invokes mathematics as a corrective: "This alone, of course, would have caused the truth to be hidden from the human race to eternity, if mathematics, which is concerned not with ends, but only with the essences and properties of figures, had not shown men another standard of truth."[49] This scientific truth, grounded in necessity, excludes normative notions like goodness or beauty, which imply intent. God's perfection, for Spinoza, lies in the unity of His essence, will, and reason—His existence inherently generates all things as an expression of power, not design. Purpose, he contends, is a human projection, and knowledge derived from imagination falls short of truth. This purposelessness constitutes true freedom: "For

[44] Ibid., p. 99.

[45] Ibid., p. 98.

[46] Ibid., pp. 7–8.

[47] Giordano Bruno, *Cause, Principle and Unity* (Cambridge University Press, 1998), pp. 37–38.

[48] Benedict De Spinoza, *The Ethics and Other Works*, trans. Edwin Curley (Princeton University Press, 1994), p. 97.

[49] Ibid., p. 111.

God alone exists only from the necessity of his nature and acts from the necessity of his nature. Therefore God alone is a free cause, q.e.d."[50]

Third, the value-neutrality of material nature. Spinoza uses *natura naturata* to convey the idea of material nature and it is the production of God's creation. Material nature for Spinoza doesn't have any ethical and aesthetic value. The value-neutrality of the material nature can be traced back to his metaphysical concept of God. "God is the efficient cause of all things which can fall under an infinite intellect"[51] For Spinoza, God's intellect doesn't have any value dimension. Intellect's creation is not motivated by values such as truth, goodness, and beauty. The creation is due to its necessity. The intellect has the feature of perfection, which will necessarily create all things in nature. However, physical nature as the product doesn't include existence as he interprets, "The essence of things produced by God does not involve existence."[52] For him, only God has existence. All material things depend on God's power for its existence. Without God, material things will cease to exist.

2.4 Kant: Aesthetic Appreciation of Nature Promotes Moral Perfection

For Kant, nature and value are separate, and nature itself is not an existence with intrinsic value. However, this does not affect his aesthetic appreciation of nature, which includes the aesthetic appreciation of natural forms and the great natural forces that exist in nature. But this aesthetic does not regard nature itself as valuable, but emphasizes the inner activity generated by thc forms of nature and the great natural forces.

Kant's account of aesthetic appreciation of nature centers on a distinctive, non-utilitarian pleasure arising from the harmony between intellect and imagination, triggered by nature's forms. This pleasure differs from two other types: the sensory enjoyment derived from agreeableness, which aligns with nature's instrumental use, and the conceptual satisfaction tied to goodness, which serves utilitarian ends. True aesthetic experience, for Kant, is disinterested and purposeless, focusing solely on form rather than the empirical nature elucidated by sciences like biology or ecology. Given his epistemological stance—that we cannot know the "thing-in-itself"—natural aesthetics emphasizes the internal dynamics of mental faculties activated by external forms, not nature's intrinsic essence. A second category, the sublime, emerges not from form but from nature's vast quantities and forces—the mathematical and dynamical sublime. Confronting these, the mind experiences a tension of pain and pleasure, stimulating reason to transcend sensory limits. The mathematical sublime prompts reason to grasp magnitudes beyond sensory comprehension, while the dynamical sublime elevates reason above sensibility's powerlessness, fostering

[50] Ibid., pp. 97–98.

[51] Ibid., p. 97.

[52] Ibid., p. 102.

moral detachment from material interests. For Kant, this aesthetic response originates not in nature as a totality but in its forms and forces engaging human cognition, locating beauty and sublimity within the mind, not the external world.

Crucially, Kant denies that natural aesthetics confers human dignity, reserving that role for morality, though he views aesthetics as a vital aid to moral development. This contrasts sharply with Emerson, who, as this study explores, grounds dignity in the unity of humanity, nature, and God. To understand the link between Kant's natural aesthetics and moral perfection, consider two key passages. First, he writes, "I do assert that to take an **immediate interest** in the beauty of **nature** (not merely to have taste in order to judge it) is always a mark of a good soul, and that if this interest is habitual, it at least indicates a disposition of the mind that is favorable to the moral feeling, if it is gladly combined with the **viewing of nature**. It must be remembered, however, that I mean here strictly the beautiful *forms* of nature, and by contrast set to one side the *charms* that it usually combines so abundantly with them, since the interest in them is to be sure also immediate, but nevertheless empirical."[53] Second, he elaborates, "Someone who alone (and without any intention of wanting to communicate his observations to others) considers the beautiful shape of a wildflower, a bird, an insect, ect., in order to marvel at it, to love it, and to be unwilling for it to be entirely absent from nature, even though some harm might come to him from it rather than there being any prospect of advantage to him from it, takes an immediate and certainly intellectual interest in the beauty of nature. I.e., not only the form of its product but also its existence pleases him, even though no sensory charm has a part in this and he does not combine any sort of end with it."[54] These passages invite analysis from two perspectives.

First, they illuminate Kant's conception of natural aesthetics as a disinterested pleasure, distinct from sensory gratification or conceptual utility. In Kant's *Critique of the Power of Judgment*, this harmony of intellect and imagination, elicited by nature's forms, eschews ulterior motives, mirroring the purity of moral will—which acts from duty, not inclination or personal ties. While Kant disregards nature's sensory qualities, focusing solely on form, this does not license its arbitrary treatment. Although he ascribes intrinsic value only to humans as ends, the non-utilitarian nature of aesthetic appreciation fosters a disposition that resists exploitation, reflecting a soul attuned to goodness. By freeing the will from sensory influences, such appreciation elevates the soul, aligning it with moral law beyond mere pleasure-seeking. Thus, Kant's anthropocentrism does not theoretically justify harming nature's creatures.

Second, these passages reveal a reciprocal affinity between aesthetic and moral sensibilities. A soul attuned to morality's goodness is primed to find beauty in nature, while aesthetic engagement reciprocally nurtures moral development. Kant further asserts, "A propensity to wanton destruction of what is *beautiful* in inanimate nature is opposed to a human being's duty to himself; for it *weakens* or uproots that feeling in him which, though not of itself moral, is still a disposition of sensibility that greatly

---

[53] Immanuel Kant, *Critique of the Power of Judgment*, trans. Paul Guyer and Eric Matthews (Cambridge University Press, 2000), p. 178.

[54] Ibid., pp. 178–179.

promotes morality or at least prepares the way for it: the disposition, namely, to love something (e.g., beautiful crystal formations, the indescribable beauty of plants) even apart from any intention to use it. With regard to the animate but nonrational part of creation, violent and cruel treatment of animals is far more intimately opposed to a human being's duty to himself, and he has a duty to refrain from this."[55] Though aesthetic emotions are not moral, they cultivate a disinterested love that bolsters reason's dominion over inclination, countering humanity's finite pursuit of happiness. This aligns with Kant's broader ethic of self-cultivation, where developing natural endowments—mental, spiritual, and physical—is a duty to oneself, exercised through freedom, not instinct, to fulfill rational purposes.

3 Nietzsche: Nature and Bestowal of Value

The final type of relationship between nature and value is grounded in Nietzsche's philosophy. For him, nature is characterized by chaos rather than by rational order. He conceives nature as manifestation of the will to power, utterly devoid of inherent truth or objective value. In this view, nature possesses no standalone significance; rather, it is the intrinsic human pursuit of meaning that actively projects and bestows value upon an otherwise indifferent natural world. Nietzsche does not use scientific methods to explain nature. He thinks that there is no objective truth, which is very different from the philosophers mentioned earlier, such as Spinoza and Descartes. According to Nietzsche, there is no so called truth. Truth for him is constructed under the guidance of the will to power. We can never reach a completely true world, and the search for such a world is just a duty and need. Nietzsche believes that objective truth does not exist. He writes, "The world with which we are concerned is false, i.e., is not a fact but a fable and approximation on the basis of a meager sum of observations; it is 'in flux,' as something in a state of becoming, as a falsehood always changing but never getting near the truth: for—there is no 'truth.'"[56] When Nietzsche says that the world is flowing, he does not mean that the essence of the world is flowing. Instead, he stresses that the world is constantly generated in our interpretation, which cannot bring us closer to truth. Interpretation is the result of human beings' will to power and they try to interpretate the world from the perspective of value. "The question of values is more *fundamental* than the question of certainty: the latter becomes serious only by presupposing that the value question has already been answered."[57]

Another manifestation of Nietzsche's metaphysical thought is the division of two worlds, opposing people's view of the fictional ideal world as the most real world. Among the philosophers we discussed in the first and second parts, most

[55] Immanuel Kant, *The Metaphysics of Morals*, trans. Mary Gregor (Cambridge University Press, 1996), p. 192.

[56] Friedrich Nietzsche, *The Will to Power*, ed. Walter Kaufmann and R. J. Holingdale (New York: Vintage Books, 1968), p. 330.

[57] Ibid., p. 322.

philosophers' metaphysics is based on two worlds. For example, Plato regards the world of perfect ideas beyond specific time and space as the real world, and the world we live in is just a shadow of the world of ideas, and does not have reality. Kant divides the world into the world of phenomena and the world of things in themselves. We can only have knowledge of the world of phenomena, and we can never gain knowledge of the world of things in themselves. For Nietzsche, there is no other world of truth apart from the human world. Because the so-called world of truth is also just an interpretation of human will to power. For example, his criticism of Descartes' thinking self, "*It* thinks; but that this 'it' is precisely the famous old 'ego' is, to put it mildly, only a supposition, an assertion, and assuredly not an 'immediate certainty.' After all, one has even gone too far with this 'it thinks'—even the 'it' contains an *interpretation* of the process, and does not belong to the process itself."[58] For Descartes, "I" refer to a pure consciousness stripped of specific content, which carries out many activities, including affirmation, denial, doubt, and so on. But Nietzsche emphasizes that there is no self-existing thing that supports thought itself. If all the relationships, characteristics, or activities of a thing are set aside, the self-existing thing itself cannot exist at all. The self-existing thing itself is actually constructed by human will to power, with the sole purpose of description and understanding.

After opposing the use of truth to explain nature, Nietzsche reinterpreted it with the concept of will to power. The concept of will to power transcends the objective truth, goodness, and beauty proposed by many Western philosophers. He emphasized that human will to power will continuously interpret nature, endow nature with value, so as to satisfy human pursuit of meaning.

First, Nietzsche employs the concept of the will to power to reinterpret nature, rejecting its depiction as an objective, material reality. Instead, he sees beneath its surface a dynamic interplay of desires—survival, competition, and other drives—as real as the will to power itself. He writes, "Suppose nothing else were 'given' as real except our world of desires and passions, and we could not get down, or up, to any other 'reality' besides the reality of our drives—for thinking is merely a relation of these drives to each other: is it not permitted to make the experiment and to ask the question whether this 'given' would not be *sufficient* for also understanding on the basis of this kind of thing the so-called mechanistic (or 'material') world? I mean, not as a deception, as 'mere appearance,' an 'idea' (in the sense of Berkeley and Schopenhauer) but as holding the same rank of reality as our affect—as a more primitive form of the world of affects in which everything still lies contained in a powerful unity before it undergoes ramifications and developments in the organic process (and, as is only fair, also becomes tenderer and weaker)—as a kind of instinctive life in which all organic functions are still synthetically intertwined along with self-regulation, assimilation, nourishment, excretion, and metabolism—as a *pre-form* of life."[59] For Nietzsche, no objective truth exists independently of human perception. For him,

---

[58] Friedrich Nietzsche, *Beyond Good and Evil*, trans. Walter Kaufmann (New York: Vintage Books, 1966), p. 24.

[59] Ibid., pp. 47–48.

ontological reality doesn't exist and this idea is the result of human beings' will to power. Humans are embedded in nature, which he views as a tangible reality—not a Platonic shadow, nor a Berkeleyan or Schopenhauerian illusion. Crucially, nature's reality mirrors that of human emotions, both rooted in a will to power manifesting as an instinctual life force. This force, more primal than mere physiological impulses, encompasses self-regulation, assimilation, nutrition, excretion, and metabolism, uniting nature and humanity without hierarchy. Nietzsche's metaphysical vision thus portrays nature as chaotic, purposeless, and value-neutral, stripped of inherent order or moral significance.

Second, Nietzsche conceives of nature as fundamentally chaotic, devoid of inherent order. He asserts, "The total character of the world, however, is in all eternity chaos—in the sense not of a lack of necessity but of a lack of order, arrangement, form, beauty, wisdom, and whatever other names there are for our aesthetic anthropomorphisms."[60] This perspective starkly contrasts with the Romantic visions of Rousseau and Emerson, revealing a profound divergence in their conceptions of the ultimate good. For Rousseau, nature, particularly the wilderness, embodies beauty and freedom; for Emerson, it serves as the universal principle, generating beauty such that all wilderness existence reflects this aesthetic ideal. Nietzsche, however, perceives nature as lacking order, form, or beauty—an existence stripped of such qualities. Crucially, this view is not an ontological deduction derived from rational analysis but a product of the will to power. Nietzsche contends that humans, unable to apprehend nature in its intrinsic form, impose chaos as an interpretive lens. He elaborates, "Let us beware of saying that there are laws in nature. There are only necessities: there is nobody who commands, nobody who obeys, nobody who trespasses. Once you know that there are no purposes, you also know that there is no accident; for it is only beside a world of purposes that the word 'accident' has meaning."[61] Here, necessity governs nature, not laws or purposes, rendering traditional notions of order illusory.

Yet, Nietzsche's characterization of nature as chaotic does not imply sheer disorder. Rather, it suggests an order emergent within chaos, a point Heidegger incisively critiques: "However, this 'chaos' must not be taken to mean a blind, turbid swirling about in confusion; rather, it means the flowing diversity of the whole of beings, a diversity that is ever borne along into a certain order of power, that limits domains of power, and that ever strives to reach decisions in the strife over the scope of power."[62] Nietzsche thus envisions nature as a diverse whole, not harmoniously ordered but dynamically engaged in power struggles, seeking provisional structures through the will to power. Consequently, nature resists value-laden descriptors such as "good" or "beautiful," remaining value-neutral. This stance reflects Nietzsche's broader metaphysics, where chaos is not an objective truth but a human valuation, shaped by interpretive acts rather than inherent properties. The contrast

---

[60] Friedrich Nietzsche, *The Gay Science*, trans. Walter Kaufmann (New York: Random House, 1974), p. 168.

[61] Ibid., p. 168.

[62] Martine Heidegger, *Nietzsche*, Volume II (Harper & Row, 1984), p. 103.

with Rousseau and Emerson underscores how Nietzsche's rejection of nature's aesthetic or moral significance reorients the pursuit of the good toward power and self-overcoming, rather than harmony or beauty.

Third, Nietzsche posits nature as not only chaotic but also devoid of purpose, a stance central to his metaphysics. He argues that the primary impulse of natural life forms is to discharge their power, an expression of the will to power, rather than to pursue self-preservation. The latter, he argues, emerges as an indirect or frequent outcome, not an inevitable one. By interrogating Spinoza's concept of self-preservation as nature's teleological principle, Nietzsche exposes a profound internal inconsistency in him: Spinoza presupposes purpose where Nietzsche denies it exists. He asserts that assigning purposes to nature stems from human constructs—social customs, political conventions, and artistic or religious traditions—rather than any inherent quality. He writes,

> And do you know what "the world" is to me? Shall I show it to you in my mirror? This world: a monster of energy, without beginning, without end; a firm, iron magnitude of force that does not grow bigger or smaller, that does not expend itself but only transforms itself; as a whole, of unalterable size, a household without expenses or losses, but likewise without increase or income; enclosed by "nothingness" as by a boundary; not something blurry or wasted, not something endlessly extended, but set in a definite space as a definite force, and not a space that might be "empty" here or there, but rather as force throughout, as a play of forces and waves of forces, at the same time one and many, increasing here and at the same time decreasing there.[63]

Here, conquest and domination—reinterpreting and reshaping prior meanings—define organic processes, erasing any original purpose.

This framework renders nature value-neutral under the will to power. Nietzsche observes, "We enjoy being in the countryside so much because it has no opinion concerning us."[64] These reflections appear contradictory: nature's lack of judgment offers respite, yet its indifference provokes restlessness. This contrasts sharply with Emerson, who elevates beauty as the universe's highest principle, rendering nature an inexhaustible source of aesthetic delight. For Nietzsche, nature's purposelessness and neutrality preclude such valorization.

Though value-neutral, nature does not exist for human ends. Humans are not the universe's value center. Nietzsche remarks, "To what end the 'world' exists, to what end 'mankind' exists, ought not to concern us at all for the moment except as objects of humour: for the presumptuousness of the little human worm is the funniest thing at present on the world's stage;"[65] Rejecting religious or transcendental views that imbue nature with purpose, goodness, or beauty, Nietzsche redefines its value as a construct of subject and object interplay, a topic I will explore through these dual perspectives.

---

[63] Friedrich Nietzsche, *The Will to Power* (Vintage Books, 1968), pp. 549–550.

[64] Friedrich Nietzsche, *Human, All Too Human*, trans. R. J. Hollingdale (Cambridge University Press, 1986), p. 181.

[65] Friedrich Nietzsche, *Untimely Meditations*, trans. R. J. Hollingdale (Cambridge Unviersity Press, 1983), p. 112.

In sum, the understanding of value by different philosophers has huge influence on people's attitude towards nature. Understanding the relationship between nature and value generates three types of relationship with nature. (1) Living in harmony with nature. If nature is regarded as the bearer of ultimate value including truth, goodness, and beauty, this vision will encourage the attitude of living in harmony with nature rather than dominating nature; If nature is regarded as separation from ultimate value, it will bring two types of attitudes. (2) Dominating nature. Nature is regarded a machine which can be better understood by physics and geometry. (3) Know nature to better know ourselves. I will examine the three attitudes as follows.

First, the attitude of living in harmony with nature. The above-mentioned philosophers such as Aristotle, Plotinus, Augustine, and Rousseau's interpretation of nature and value impact an attitude of living in harmony with nature. They all give strong objection towards dominating attitude towards nature. Rousseau and Plotinus, for example, has explicitly used the word harmony. Rousseau states, "Let us return to nature, which never lies, and live in harmony with her laws, which are the laws of our own being."[66] Plotinus says, "The Universe is a thing of beauty and its parts are in harmony with one another, because it is the product of the soul, which imposes unity and coherence upon it…Those who seek to dominate or exploit it fail to see its true nature, for they are blinded by their own desires."[67] From the above quotes, it can be seen that the deep philosophical reason why they object to the dominating attitudes towards nature is the ultimate value in nature. Nature has its own nature and laws which are the foundation for human nature. Aristotle also objects to dominating attitudes towards nature. Many scholars such as David Sedley[68] and Christopher V. Mirus[69] have pointed out that for Aristotle, there exists an order in nature, human beings' end should align with nature's ends and in this way, humanity can achieve ultimate goodness. In fact, Aristotle's understanding of human nature is based on his understanding of nature. In this aspect, it is similar to Plotinus and Rousseau.

Second, the attitude of dominating nature. Descartes separates value from nature, which brings the dominating attitudes towards nature. Although Descartes stresses truth, his concept of truth is not value-laden rather, it is certainty as understood by science. Certainty for him is for the use of domination over nature as he explicitly expresses, "For these notions made me see that it is possible to arrive at knowledge that would be very useful in life and that, in place of that speculative philosophy taught in the schools, it is possible to find a practial philosophy, by means of which, knowing the force and the actions of fire, water, air, the stars, the heavens, and all the other bodies that surround us, just as distinctly as we know the various skills of our craftsmen, we might be able, in the same way, to use them for all the purposes for which they are appropriate, and thus render ourselves, as it were, masters and

---

[66] Jean-Jacques Rousseau, *The Social Contract and Discourses*, trans. G.D.H. Cole (London: J. M. Dent 1973), p. 84.

[67] Plotnius, *The Enneads*, trans. Stephen Mackenna (Penguin Books, 1991), p. 92.

[68] David Sedley, "Is Aristotle's Teleology Anthropocengric?", Phronesis 36, no. 2 (1991): 179–196.

[69] Christopher V. Mirus, "Aristotle on Beauty and Goodness in Nature," Intentional Philosophical Quarterly 52, no. 1 (2012): 79–97.

possessors of nature."[70] For Descartes, the motivation for us to know nature is to dominate it to serve the human beings' welfare. For him, to know nature for its own sake doesn't have any meaning.

Third, the attitude of knowing nature for the reason of better knowing ourselves. Although many modern philosophers separate value from nature, it doesn't necessarily lead to the domination of nature. For example, Spinoza and Kant, both think the nature itself doesn't have value for its own sake. Nature itself for them doesn't have ethical and aesthetic value. For example, Kant writes, "The beautiful in nature is not beautiful in itself as it might seem to be for an empirical judgement; rather, it is beautiful only through the form it assumes in the mind's presentation of it."[71] For Kant, beauty doesn't exist in nature, rather, it is the interaction between the form of nature and human beings' minds. Nature for Kant also doesn't have ethical value since in this world, the only thing that has value for its own sake is human beings' good will. Value lies in human beings' goodwill. However, Kant stresses that "A propensity to wanton destruction of what is *beautiful* in inanimate nature is opposed to a human beings' duty to himself; for it weakens or uproots that feeling in him which, though not itself moral, is still a disposition of sensibility that greatly promotes morality or at least prepares the way for it: the disposition, namely, to love something (e.d., beautiful crystal formations, the indescribable beauty of plants) even apart from any intention to use it."[72] According to Kant, appreciation of the beauty of nature can strengthen humanity's effort of attaining virtue. Spinoza, like Kant, also objects to the domination of nature. The deeper philosophical reason lies in his metaphysics, which generates intellectual humility in human beings. For him, "The force by which a man perseveres in existing is limited, and infinitely surpassed by the power of external causes."[73] For Spinoza, human beings' metaphysical existence depends on nature's power, which is governed by reason. If human beings want to flourish, they must know nature intellectually so that they can align with nature's power. From the above discussion, it can be seen that there is a general trend that the philosophical importance of material nature has been neglected by many philosophers. In the first type of nature and value relationship, as I examined above, although nature is regarded as being the value bearer, material nature is still less important than metaphysical nature. For example, Aristotle regards form as being more important. For him, the good life is the contemplation of unmoved mover, God. He will not attach to a certain place such as landscape and wilderness, and aesthetically appreciate it as Emerson always did. For Augustine, The beauty of God is superior to the beauty of all material nature. He stresses our contemplation of God rather than our contemplation of material objects. Within this type, Plotinus and Rousseau are

---

[70] René Descartes, *Discourse on the Method and Meditations on First Philosophy*, trans. Donald A. Cress (Hackett Publishing Company, 1998), p. 35.

[71] Immanuel Kant, *Critique of Judgment*, trans. Werner S. Plubar (Indianapolis: Hackett Publishing, 1987), p. 45.

[72] Immanuel Kant, *The Metaphysics of Morals*, trans. Mary Gregor (Cambridge University Press, 1996), p. 192.

[73] Benedict De Spinoza, *A Spinoza Reader: The Ethics and Other Works*, trans. Edwin Curley (Princeton University Press, 1994), p. 202.

the exceptions. Their metaphysical interpretation of nature lays the foundation for aesthetically appreciate material nature as Emerson. However, Emerson develops their ideas and brings more science into his interpretation of material nature. So the aesthetic appreciation of it is enriched by science. In the second type of nature and value relationship, we can find Descartes, Spinoza, and Epicurus all neglect aesthetic appreciation of material nature. For Nietzsche, although his concept of will to power brings the new understanding of nature, he rejects the thought that nature has value for their own sake. Nature is in essence a will to power.

References

1. Aristotle, *Physics*, trans. Robin Waterfield (Oxford University Press, 1996).
2. Aristotle, *Parts of Animals*, trans. A. L. Peck (Cambridge, Massachusetts, 1961).
3. Plotinus, *The Ennead*, trans. Stephen MacKenna (Penguin Books, 1991).
4. Aristotle, *History of Animals*, trans. Wu Shoupen (The Commercial Press 2010).
5. Plotinus, *The Ennead*, trans. Stephen MacKenna (Penguin Books, 1991).
6. Augustine, *Confessions*, trans. Henry Chadwick (Oxford University Press, 1991).
7. Augustine, *The City of God*, Volume 1, ed. Marcus Dods (T&T Clark, 1871), p. 485.
8. Jean-Jacques Rousseau, *Emile or On Education*, trans. Allan Bloom (New York: Basic Books, Inc., Publishers, 1979).
9. Jean-Jacques Rousseau, *Reveries of the Solitary Walker*, trans. Peter France (London: Penguin Classics, 2004).
10. *Great Books of the Western World* (12), ed. Robert Maynard Hutchins, (EncyclopÆdia Britannica Inc. 1952).
11. *Epicurus: The Extant Remains*, trans. Cyril Bailey, M.A. (Oxford: At The Clarendon Press, 1926).
12. René Descartes, *Meditations on First Philosophy*, trans. Donald A. Cress (Hackett Publishing, 1998).
13. Benedict De Spinoza, *The Ethics and Other Works*, trans. Edwin Curley (Princeton University Press 1994).
14. Giordano Bruno, Cause, *Principle and Unity* (Cambridge University Press, 1998), pp. 37–38.
15. Immanuel Kant, *Critique of Judgment*, trans. Paul Guyer, Eric Matthews (Cambridge University Press, 2000), p. 178.
16. Ibid., p. 178.
17. Immanuel Kant, *The Metaphysics of Morals*, trans. Mary Gregor (Cambridge University Press, 1996).
18. Friedrich Nietzsche, *The Will to Power*, ed. Walter Kaufmann (New York: Vintage Books, 1968).
19. Friedrich Nietzsche, *Beyond Good and Evil*, trans. Walter Kaufmann (New York: Vintage Books, 1989).
20. Friedrich Nietzsche, *The Gay Science*, trans. Walter Kaufmann (New York: Random House, 1974).
21. Martine Heidegger, *Nietzsche*, Volume II (Harper & Row, 1984).
22. Friedrich Nietzsche, *Human, All Too Human*, trans. R. J. Hollingdale (Cambridge University Press, 1986).
23. Friedrich Nietzsche, *Untimely Meditations*, trans. R. J. Hollingdale (Cambridge Unviersity Press, 1983).
24. Jean-Jacques Rousseau, *The Social Contract and Discourses*, trans. G.D.H. Cole (London: J. M. Dent 1973).

25. Plotnius, *The Enneads*, trans. Stephen Mackenna (Penguin Books, 1991).
26. David Sedley, "Is Aristotle's Teleology Anthropocentric?", Phronesis 36, no. 2 (1991).
27. Christopher V. Mirus, "Aristotle on Beauty and Goodness in Nature," Intentional Philosophical Quarterly 52, no. 1 (2012).
28. René Descartes, *Discourse on the Method and Meditations on First Philosophy*, trans. Donald A. Cress (Hackett Publishing Company, 1998).
29. Immanuel Kant, *Critique of Judgment*, trans. Werner S. Plubar (Indianapolis: Hackett Publishing, 1987), p. 45.
30. Immanuel Kant, *The Metaphysics of Morals*, trans. Mary Gregor (Cambridge University Press, 1996).
31. Aristotle, *Metaphysica in The Works of Aristotle*, vol. VIII, trans. W. D. Ross (Oxford at the Clarendon Press, 1928).

Chapter 2
Spirituality

The meaning of spirituality is originally associated with religion, which is mostly directed to ethical life. And it is gradually extended to meanings and values in people's everyday life. In the past thirty years, philosophers have given a deep philosophical analysis of the concept of spirituality. Many works have been published in this field. These works include David Carr and John Haldane's *Spirituality, Philosophy, and Education* (2003), John Cottingham's *The Spiritual Dimension: Religion, Philosophy and Human Value* (2005), David McPherson's *Spirituality and the Good Life* (2017), and Robert C. Solomon's *Spirituality for Skeptics: The Thoughtful Love of Life* (2002). Although these works have taken different approaches and provided different philosophical interpretations of this concept, they share some similar viewpoints in their interpretation. For them, spirituality includes the following dimensions: (1) Meaning in life and a good life. Spirituality is firstly related to the pursuit of good life. The good life is also the pursuit of meaning in life. In terms of meaning in life, various philosophical theories have been constructed to this topic which include objective, subjective, hybrid, and supernatural theory; (2) Self-transcendence. It contains the inner transformation, that is, the transformation from the false self to the true self. John Cottingham interprets transformation as "radical reorientation of the self"; (3) Value. Spirituality includes the pursuit of ultimate values which are various expressions of truth, goodness, and beauty; Some philosophers stress the moral values and others stress the aesthetic values; Some will stress combination of truth, goodness, and beauty; (4) Spiritual practices. For this dimension, John Cottingham terms it as "praxis" component of spirituality. Spirituality contains various kinds of practices which can help people to achieve self-transcendence. These practices include rational dialogue, daily routines, retreat, pray, singing of Psalms, meditation, communal living, self-reflection, performance of various rites, chanting, fasting, and so on.

S. Gao, *Comparative Studies Between Emerson and Chinese Philosophy*, Nature Aesthetics: Interdisciplinary and Intercultural Dialogue,
https://doi.org/10.1007/978-981-95-0421-3_2

1 The Pursuit of Meaning in Life

Spirituality includes the recognition that diverse ways of living exist. Some of the ways of living will be considered as the good life with meaning. In the past decades, philosophers have developed four types of theories in terms of meaning in life.

First, objective theories. Thaddeus Metz champions this approach, arguing that meaning derives from objective values independent of divine or supernatural entities. Critiquing Susan Wolf's subjective emphasis on love for one's pursuits, Metz asserts, "A human person's is more meaningful, the more that she, without violating certain moral constrains against degrading sacrifice, employs her reason and in ways that either positively orient rationality towards fundamental conditions of human existence, or negatively orient it towards what threatens them, such that the worse parts of her life cause better parts towards its end by a process that makes for a complelling and ideally original life-story; in addition, the meaning in a human person's life is reduced, the more it is negatively oriented towards fundamental conditions of human existence or exhibits narrative disvalue."[1] Metz posits that truth, goodness, and beauty possess intrinsic value, broader than narrower foci on single aspects. Meaning arises not from enjoyment but from rational engagement with human fundamentals, requiring rigorous effort and planning;

Second, subjective theories. The representative for subjective theory in the meaning of life is Richard Taylor. He uses the example of ancient myth of Sisyphus to illustrate that life's meaning lies in the subjective feelings of love and passion towards doing the things even if the things people are doing are meaningless for most people. He argues that "The meaning of life is from within us, it is not bestowed from without, and it far exceeds in both its beauty and permanence any heaven of which men have ever dreamed or yearned for."[2] For Richard, people can impose meaning to the things they are doing.

Third, hybrid theories. Susan Wolf is the representative of these theories. Susan Wolf integrates subjective and objective elements and gives a definition of meaningfulness. She writes, "According to the conception of meaningfulness I wish to propose, meaning arises from loving objects worthy of love and engaging with them in a positive way."[3] Drawing on Aristotle's experiential method in *Nicomachean Ethics*, Wolf bases her theory on common sentiments: "Do what you love" and "Find something worth loving." Meaning requires active participation in beloved pursuits for self-fulfillment, yet love alone suffices only if directed toward objectively worthy projects.

Fourth, supernaturalist theories. By supernaturalism, I refer to the transcendental realities which include transcendental God or some transcendental values. Joseph Mintoff is one of the representative of this theory. Joseph Mintoff posits that meaning

---

[1] Thaddeus Metz, *Meaning in Life* (Oxford University Press, 2013), p. 235.

[2] Richard Taylor, "The Meaning of Life" in *Good and Evil* (Prometheus Books, 2000), pp. 319–334.

[3] Suan Wolf, *Meaning in Life and Why It Matters* (Princeton University Press, 2010), p. 8.

hinges on transcendent entities beyond individual life. He argues, "One's life is meaningful to the extent it involves attachments to transcendent things."[4] These cosmic realities transcend mundane human values such as freedom, or health, functioning instead as the ultimate ends. Distinguishing moral action from cognition of goodness, he adds, "Indeed, the exercise of moral virtue is not even necessary for a meaningful life."[5] For him, knowing good is more meaningful than doing good for moral cause.

The good life, as understood by Emerson and Chinese philosophy, which are the focus of this book, are both related to nature. They both regard a life in harmony with nature as a good life and a life with meaning. Emerson has expressed clearly in his work that nature represents the best meaning of life. In Confucianism and Daoism, although the concept of meaning doesn't appear in their work, scholars have reached a consensus that living in harmony with Heaven and Earth is the dominating spirit of Chinese philosophy. Both Emerson and Chinese philosophy do not establish a theory of the meaning of life, but if we classify Emerson and Chinese philosophy according to the above types in order to better understand their philosophy. Emerson can be classified as the supernatural theory of the meaning of life. Although Emerson doesn't neglect the mundane life, his final purpose is still to know God and united with God. Confucianism and Daoism in Chinese philosophy can be classified as the hybrid theories of the meaning of life. Both Confucianism and Daoism emphasize human virtue which demonstrate the value of good. They also stress human beings' love the virtue. Both Confucianism and Daoism don't belong to supernatural theory. For the two schools of philosophy, everyday life is more important than being connected to transcendental entities. For them, doing good and right thing is most important than knowing good.

2 Value

Value represents the second key dimension of spirituality. Krzysztof Piotr Skowroński in his book titled *Values and Powers: Re-reading the philosophical tradition of American pragmatics*[6] traces the study of value in Western philosophical tradition. He provides a list of important philosophical works in value inquiry, which include Christian von Ehrenfels' *System of Theory of Value* (1893), G. E. Moore's Principia Ethica (1903), John Laird's *The Idea of Value* (1920). Nietzsche's provocative call for a "re-evaluation of all values" invigorated this discourse, challenging conventional frameworks and inspiring new approaches. John Dewey, in his book titled *Experience and Nature* (1925), reconceived value as arising from lived experience, critiquing the metaphysical foundations of traditional Western philosophy and grounding value in human practice. Krzysztof Piotr Skowronński enriches the study

[4] Joe Mintoff, "Transcending Absurdity", *Ratio* XXI 1, p. 81.

[5] Ibid., p. 71.

[6] Krzysztof Piotr Skowroński, *Values and Powers: Re-reading the Philosophical Traditional of American Pragmatism* (Amsterdam-New York, NY 2009), pp. 11–13.

of values with an interdisciplinary approach, delineating ten perspectives—ontology, epistemology, anthropology, sociology, psychology, typology, deconstruction, hierarchy, deontology, and methodology—that illuminate value's multifaceted nature and invite deeper philosophical reflection.

This book, however, departs from both a narrowly disciplinary treatment of value philosophy and Skowroński's broad interdisciplinary perspectives. Instead, it engages value through core philosophical concerns germane to its central theme: a comparative exploration of spirituality, nature, and place in Emerson and Chinese philosophy. The analysis centers on four interrelated dimensions: (1) the relationship between value and obligation, probing their mutual influence; (2) the transcendental and experiential aspects of value, examining its simultaneous immanence and elevation beyond the mundane; (3) the nexus of value and power, exploring their dynamic interplay. These dimensions frame the subsequent discussion, rooting the concept of value within the book's broader spiritual and philosophical inquiry; (4) Spiritual practices. Spiritual practices take different forms including meditation, prayer, rational dialogue.

2.1 Values and Obligation

The relationship between value and obligation is often framed in the central philosophical question: does acknowledging an entity's value impose a duty to respect it? Immanuel Kant answers affirmatively, asserting that the good will uniquely possesses unconditional value, serving as the foundation for our moral obligations towards human beings. He writes, "There is nothing it is possible to think of anywhere in the world, or indeed anything at all outside it, that can be held to be good without limitation, excepting only a good will."[7] This absolute value derives from its autonomous capacity to legislate and enact moral law, independent of external outcomes. For Kant, we must respect people and promote their happiness for the reason that they have intrinsic value. Modern environmental philosophers extend this Kantian framework beyond humanity. Tom Regan, a proponent of animal rights, applies deontological principles to animals, arguing they possess intrinsic value as "subjects-of-a-life." He states, "... individuals are subjects-of-a-life if they have beliefs and desires; perception, memory, and a sense of future, including their own future; an emotional life together with feelings of pleasure and pain; preference and welfare-interests; the ability to initiate action in pursuit of their desires and goals; a psycho-physical identity over time; and an individual welfare in the sense that their experiental life fares well or ill for them, logically independently of their utility for others and logically independently of their being the object of anyone else's interests."[8] For Regan, if human beings rationally justify this value, we will have a direct moral duty to respect

[7] Immanuel Kant, *Groundwork of the Metaphysics of Morals*, trans. Allen W. Wood (Yale University Press, 2002), p. 9.

[8] Tom Regan, *The Case for Animal Rights* (University of California Press, 1983), p. 243.

and protect these animals. Holmes Rolston III broadens this concept of value to encompass wilderness, asserting its intrinsic value justifies wilderness preservation. He argues wilderness is full of creative power and it is always in the process of generating new lives. It has intrinsic value for its own sake rather than for people's use. Human beings have direct obligation towards wilderness. He extends the concept of intrinsic value in Kant to nature. For him, although human beings have the highest moral status, wilderness also has moral status since it has intrinsic value.[9] Regan and Rolston's justification for preservation of animals or wilderness is in essence the extension of Kant's idea of absolute and intrinsic value of human beings.

Eugene C. Hargrove integrates aesthetics into this ethical paradigm. For him, aesthetics is more important than ethics. The beauty of nature has value and it is a good. He argues, "In addition, the duty specifically to promote and preserve natural beauty arises out of the recognition that not only artistic beauty but also natural beauty constitutes an aesthetic good that makes up part of the general good that exists and ought to exist in the world."[10] This aesthetic-ethical linkage finds practical expression in conservation policy, as Victoria Jenkins illustrates: "It argues that natural beauty can form the basis of a reasoned, robust and transparent process of designation that can account for all the values of a protected landscape – as nature, culture and in terms of its beauty."[11]

The interplay between value and obligation is bound up with the hierarchy of values, which comprises both distinctions among different value domains and the internal gradations that arise within any one of them, whether in the sphere of scientific truth, moral goodness or beauty. Among diverse values, debates persist: does beauty outrank goodness? Some scholars prioritize aesthetic pursuit, elevating beauty above ethical considerations. Within a given value paradigm such as beauty, internal taxonomies emerge through the distinction between artistic and natural beauty, thereby complicating questions of their relative priority. Hargrove argues against most philosophers who maintain that natural beauty is aesthetically inferior to artistic beauty.[12] For Hargrove, most of natural beauty is superior to the beauty of art. Rolston develops the position that hierarchies exist within moral goodness. He thinks that organisms, wildlife, species, and ecosystems all possess intrinsic value, yet their worth varies by scale. The ecosystem's value is paramount, integrating lesser entities into a systemic whole. Thus, the obligation to protect wilderness supersedes that of individual organisms, reflecting the priority of ecological integrity.[13] Philosophical discussion of hierarchy of values is very important in policy making and protection practices. It directly determines our obligation towards valuable things.

---

[9] Holmes Rolston III, *Environmental Ethics: Duties to and Values in the Natural World* (Temple University Press, 1988).

[10] Eugene C. Hargrove, *Foundations of Environmental Ethics* (Prentice Hall, 1989), p. 192.

[11] Victoria Jenkins, In Defense of Natural Beauty: Aesthetic Obligation and the Law on the Designation of Protected Landscapes in England and Wales, *Environmental Law Review*, Volume 22, Issue 1, 2020, pp. 7–24

[12] Eugene C. Hargrove, *Foundations of Environmental Ethics* (Prentice Hall, 1989), p. 192.

[13] Holmes Rolston III, *Environmental Ethics: Duties to and Values in the Natural World* (Temple University Press, 1988).

In Emerson and Chinese philosophy, nature is closely connected to value. However, it doesn't have the intrinsic value as Holmes Rolston III and Hargrove have claimed. In Emerson, God has the ultimate value and we have direct more obligation towards God. In Chinese philosophy, our obligation is directed towards human beings which are situated in relationship with nature. When I make the comparative studies, I will have this dimension of value as the background of analysis.

2.2 *Transcendental and Experiential Aspects of Values*

The nature of value marks a fundamental divide in philosophical thought, reflecting divergent theoretical orientations. Plato exemplifies the transcendent view, positing that values such as the truth, the goodness, and the beauty possess an ontological existence independent of the material world. He asserts, "There exists an eternal, unchangeable realm of Forms, where the Good itself resides, beyond all becoming, perfect and indestructible, the cause of all that is right and beautiful in the sensible world."[14] These values, eternal and devoid of internal contradictions, transcend empirical contingency. Immanuel Kant similarly upholds transcendence, though primarily in the moral domain rather than aesthetics. He argues that the ultimate good arises innately from practical reason, not empirical discovery. This value, rooted in reason's autonomy, exists beyond the sensory realm.

In contrast, an empirical orientation locates value within the lived world. Aristotle, challenging his teacher Plato, situates happiness as the ultimate good in human practice rather than a metaphysical plane. This teleological good emerges through observable human flourishing. John Stuart Mill's utilitarianism further embeds value in experience, identifying happiness as the sole intrinsic good. John Dewey reinforces this empirical stance, rejecting metaphysical abstractions in favor of lived activity. He contends, "Of immediate values as such, values which occur and which are possessed and enjoyed, there is no theory at all; they just occur, are enjoyed, possessed; and that is all. The moment we begin to discourse these values, to define and generalize, to make distinctions in kinds, we are passing beyond value-objects themselves; we are entering, even if only blindly, upon an inquiry into causal antecedents and causative consequents, with a view to appraising the 'real,' that is the eventual goodness of the thing in question."[15] In Dewey's pragmatist framework, values are not transcendent, abstract, or eternal ideas existing independently of human activity. Instead, they are grounded in concrete experiences within the natural and social environment. Emerson is regarded as the starting point in the genealogy of pragmatism. He stresses the transcendence of ultimate values of truth, goodness, and beauty. However,

---

[14] Plato, *Republic*, trans. G. M. A. Grube (Hackett Publishing, 1992), pp. 188–189.

[15] John Dewey, *Experience and Nature* (W. W. Norton & Company, 1929), p. 403.

he also emphasizes experiential aspects of value. He states in his essay titled Experience, "Nature hates calculators; her methods are saltatory and impulsive."[16] For Emerson, the source of value lies in transcendental realities which he termed as Oversoul or God, but we can experience this value through interaction with nature. The interaction is combination of scientific, poetic, philosophical and religious. In Chinese philosophy, Dao as the value term is based on experiences instead of transcendental realities. It can be seen from the following passages in *The Analects* and *Zhuangzi*. "The Master said, 'It is Destiny if the Way prevails; it is equally Destiny if the Way falls into disuse. What can Kung-po Liao do in defiance of Destiny?"[17] For Confucius, the Way will not necessarily prevail in the world and it is out of our control. It is based on experiences, which is termed as destiny. Destiny in Confucianism is not ultimate reality. Rather it is combination of all experiences in this world. Zhuangzi also stresses the dimension of experience in the concept of Dao. In Zhuangzi, a dialogue between Master Tung-kuo and Zhuangzi can best illustrate this dimension. Master Tung-kuo askes Zhuangzi where Dao exists. Zhuangzi answers, "But you must not expect to find the Way in any particular place—there is no thing that escapes its presence!"[18] For Zhuangzi, Dao is not transcendental reality. But it is immanent in all things in the world as we experience in this world.

2.3 *Values and Powers*

Values also has a close relationship with powers. In terms of the relationship, Krzysztof Piotr Skowrosk has a deep insight. He says, "To sum up, the connection between values and powers is as follows: valuable things can be achieved by the activation of human powers, and the realization of good things needs this power. On the other hand, human energy needs valuable aims, because without valuable aims these energies would become dissipated and/or simply ill-used."[19] Skowroński identifies a dual dynamic: (1) value realization requires the mobilization of human power, and (2) human energy demands valuable goals to avoid squandering or misuse. However, I would like to point out that his explanation neglects the important metaphysical dimension of values. In this book, both metaphysical and pragmatical dimensions of values are closely related. The former is the support for the latter. Metaphysically, the logical relationship between value and power is as follows: (1) ultimate value underpins the power of ultimate facts and the human mind—without it, neither facts nor mental potency endure; (2) power shapes value, as the mind's capacity governs value

[16] Ralph Waldo Emerson, *The Portable Emerson*, ed. Jeffrey S. Cramer (Penguin Books, 2014), p. 269.

[17] Confucius, *The Analects*, trans. D. C. Lau (Penguin Books, 1979), p. 130.

[18] Zhuangzi, *Zhuangzi: The Complete Works of Zhuangzi*, trans. Burton Watson (Columbia University Press, 1968), p. 241.

[19] Kyzysztof Piotr Skowroński, *Values and Powers: Re-reading the Philosophical Tradition of American Pragmatism* (Brill, 2009), p. 11.

selection. This book privileges ultimate values in an ontological sense, viewing it as the foundational explanation of all worth. For Emerson, this ultimate value resides in God, unifying truth, goodness, and beauty. Among these three values, goodness is most important. It is the foundation for the other two values: truth and beauty. Without goodness, human beings and nature will not exist. In Chinese philosophy, ultimate value of goodness manifests through Heaven and Earth, which serve as the guarantors of human worth and the generative power sustaining the cosmos. The generative power is intrinsically intertwined with moral value of goodness.

Both traditions emphasize a creative life force emergent from ultimate value. Emerson ties this to truth's productive capacity. He invokes contemporary scientific "force" to articulate this dynamic. Chinese philosophy roots life force in the interplay of Yin and Yang, "The ten thousand things carry the yin and embrace the yang, and through the blending of the material force (*ch'i*) they achieve harmony."[20] Unlike Emerson's scientific inflection, *qi* as both the material and spiritual force underpins the process. In both, ultimate value and power converge to animate existence.

3 Self-Transcendence

Self-transcendence is the core concept of many disciplines, especially psychology, theology, and philosophy. The famous psychologist Viktor E. Frankl once offered an interpretation of this concept in his article "Self-transcendence as a Human Phenomenon."[21] This is an interdisciplinary work, and his research aims at on enhancing the happiness of the elderly. Self-transcendence is also the concept of theology. In the book titled *Power of Self-transcendence: An Introduction to the Philosophical Theology of Paul Tillich*, Guyton B. Hammond is devoted to examination of the key concept of self-transcendence in the theologian Paul Tillich's ideas. He states, "In his latest work, Tillich suggests a distinction between horizontal and vertical self-transcendence. The former refers to any movement from finite potentiality to finite actuality; It remains within the finite natural order. The latter type of self-transcendence is a movement beyond the finite world in the direction of the 'in-finite'; it is the transcendence of finitude."[22] He stresses that Tillich's concept of self-transcendence is deeply connected to life. He points out that although plants such as acorn also have self-transcendence, they don't have self-awareness like human beings. Self-transcendence in human beings is more unique and is closely related to freedom and self-awareness.

---

[20] *A Source Book in Chinese Philosophy*, trans. Wing-Tsit Chan (Princeton University Press, 1963), p. 160.

[21] Viktor E. Frankl, "Self-transcendence As A Human Phenomenon", *Journal of Humanistic Psychology*, 1966, 6 (2), pp. 97–106.

[22] Guyton B. Hammond, *Power of Self-Transcendence: An Introduction to the Philosophical Theology of Paul Tillich* (The Bethany Press, 1966), p. 37.

In the past decades, with the fast development of AI and technologies, philosophers have paid more attention to the concept and conducted a deep philosophical analysis. In philosophical research on the concept of self-transcendence, scholars usually take three different approaches. This first approaches is to analyze the concept of transcendence. The second approach is to discuss this concept in the study of the meaning of life, and the last is to combine the concepts of virtue and happiness in ethics for interdisciplinary research, such as Jennifer A. Grey and Candace Vogler's *Self-transcendence and Virtue: Perspectives from Philosophy, Psychology, and Theology.*[23] However, virtue is a manifestation of human pursuit of self-transcendence, but it is only one type of self-transcendence. Humans will also pursue self-transcendence in other values such as aesthetic values. By self-transcendence, I refer to transcending the false self, realizing the true self, and united with ultimate values. The true self is the self that represents human dignity and the most essential of being human. This self is opposed to the false self and is related to value. The interpretation of the trueself by philosophers of China and West can be divided into four types: Moral self; Cognitive self; Aesthetic self; Ecological self. Self-transcendence for the philosophers refers to achieving the true self while overcoming the false self and uniting with ultimate values.

First, the moral self and self-transcendence. Many philosophers hold that moral self represents the essence of self and human dignity. The representative philosopher for this view is Kant. For Kant, moral self represents the dignity of human beings. He writes at the beginning of his book *Groundwork of the Metaphysics of Morals*, "The good will is good not through what it effects or accomplishes, not through its efficacy for attaining any intended end, but only through its willing, i.e., good in itself, and considered for itself, without comparison, it is to be estimated far higher than anything that could be brought about by it in favor of any inclination, or indeed, if you prefer, of the sum of all inclinations."[24] Kant stresses the moral dimension of the essence of self. Although he also mentions human cognitive rationality and aesthetic judgment in his works, these two abilities do not represent human dignity. The ultimate value corresponding to the moral self is the supreme good, which is the object of human practical reason. Kant's moral self is opposed to the false self, and this false self is the radical evil of humans. The main manifestation of this fundamental evil is the reversal of order, that is, humans put the instinctive pursuit of happiness above the moral laws established by practical reason. Confucianism in Chinese philosophy also stresses the moral dimension of the self. In Confucian classics titled the *Doctrine of the Mean*, this point is clearly stated, "Humanity (*jen*) is [the distinguishing characteristic of man, and the greatest application of it is in being affectionate toward relatives."[25] Humanity in Chinese is 仁. it is the highest

[23] Jennifer A. Frey and Candace Vogler (Eds), *Self-transcendence and Virtue: Perspectives from Philosophy, Psychology, and Theology* (Routledge, 2018).

[24] Immanuel Kant, *Groundwork of the Metaphysics of Morals*, trans. Allen W. Wood (Yale University Press, 2002), p. 10.

[25] *A Source Book in Chinese Philosophy*, trans. Wing-tsit Chan (Princeton University Press, 1963), p. 104.

state of virtue that human beings should try to achieve. For Confucianism, the moral relationship with people, especially people in close relationships, defines who we are. In those relationships, we should try to fulfil our moral obligations. The primary obligation is towards our parents. In the *Analects*, there are many passages about detailed actions towards fulfilling our filial piety towards parents. "Tzu-yu asked about being filial. The Master said, 'Nowadays for a man to be filial means no more than that he is able to provide his parents with food. Even hounds and horses are, in some way, provided with food. If a man shows no reverence, where is the difference'",[26] "Tzu-hsia asked about being filial. The Master said, 'What is difficult to manage is the expression on one's face. As for the young taking on the burden when there is work to be done or letting the old enjoy the wine and the food when these are available, that hardly deserves to be called filial."[27] For Confucianism, the false self is not termed as evil like Kant has done. Instead, it is the self which is deviated from morality.

Second, the cognitive self and self-transcendence. The cognitive self corresponds to truth, but truth is not regarded as a value. Knowing the truth is stressed since it can promote welfare for human beings. Descartes and Spinoza are typical representatives. Descartes believes that this cognitive self in "I think, therefore I am" is the only certain and unquestionable existence, and the main characteristics of this self are understanding, reasoning, doubting, affirming, denying, feeling, and imagining, etc. Descartes emphasizes that human rationality can help people accurately recognize the truth. He uses wax as an example. When the wax melts and changes its shape, people still know that it is the same substance. The cognitive self emphasized by Descartes is introspective, and the process of introspection is also the process of gradually removing errors. This introspection is not about whether people have fulfilled their moral obligations, but about which of the so-called truths accepted in the mind are problematic and worthy of doubt. For Descartes, an important sign of a good life is how many errors are removed every day and how many truths are obtained. This kind of good life has ethical value, that is, goodness. The false self in Descartes is manifested when the will exceeds its own cognitive ability and extends it to things beyond the scope of the will's understanding, it will make mistakes.

Spinoza also emphasizes the cognitive self. He regards human rationality as the ability to understand for the sake of understanding, which is a manifestation of self-preservation. This understanding is not for a specific goal, but just to understand for the sake of understanding. This cognitive self also corresponds to truth itself. The concept of truth in Spinoza is not a value term. Spinoza emphasizes that reason understands the truth for the sake of understanding, and there is no purpose beyond understanding. This rational ability to understand for the sake of understanding is the ability to preserve oneself that God has given to each individual. For Spinoza, this self-preservation does not have ethical goodness, but things that promote self-preservation have ethical supreme goodness. He said that the supreme goodness of understanding is the knowledge that the human mind is consistent with nature. In

[26] Confucius, *The Analects*, trans. D. C Lau (Penguin Books, 1979), p. 64.

[27] Ibid.

Spinoza, many emotions formed in secular society (hope, fear, and pride, etc.) are representatives of the false self, and the key to restraining this false self lies in the true ideas in the mind.

Third, aesthetic self and self-transcendence. The third dimension of spirituality elevates the aesthetic self. The above mentioned moral-centric conception of self has been challenged by philosophers who critique its sufficiency for life's meaning, advocating an aesthetic pursuit. Susan wolf and Anthony Cross are the representatives for advocating aesthetic self. Susan Wolf questions morality's dominance, arguing that "If we pay increased attention and give increased weight to people's interest in getting and sustaining meaning in their lives, morality and the importance of obedience to its requirements will necessarily occupy a corresponding smaller place in our practical and evaluational outlooks. But it is arguable that the goals of morality will be as likely or more likely to be achieved, and in a way that is more rewarding to the people who are achieving them, for they will be doing so not out of obedience to duty but out of love."[28] She stresses that many aesthetic pursuits of art, the beauty of nature, or intellectual discovery are more valuable than morality since these pursuits are conducted out of love. Anthony Cross also stresses the aesthetic pursuit through aesthetic obligations directed toward objects—be they persons, artworks, or nature. He argues, "An aesthetic commitment is an active commitment made with respect to some aesthetic object. I understand the term 'object' here very broadly—the category includes discrete objects such as artworks, but it also includes aesthetic genres or categories, particular traditions or practices, and perhaps even certain more general aesthetic concepts like 'art' or 'beauty'. What distinguishes aesthetic commitments is the fact that the end to which one is committed involves substantive engagement with some aesthetic object or other."[29] This elevation of the aesthetic self finds robust precedent in thinkers who supplant morality with aesthetics. Friedrich Nietzsche rejects moral obligation as life's core, exalting the aesthetic self as the pinnacle of human potential. Oscar Wilde similarly privileges aesthetics, asserting self-realization through beauty. For Nietzsche and Wilde, the aesthetic self, liberated from moral shackles, achieves its highest expression through creative and beautiful engagement, eclipsing the moral self's limited horizon.

Fourthly, the ecological self and self-transcendence. Contemporary environmental philosophers develop this type of self. Arne Næss is the representative for promoting ecological self. Central to deep ecology is the principle of self-realization, through which individuals achieve The concept of self-transcendence is also a state of being that transcends narrow individualism. According to Næss, self-realization encompasses three core tenets: (1) All living beings possess an inherent right to pursue their own self-realization; (2) Human self-realization is inseparable from the flourishing of diverse organisms; (3) The ultimate purpose of self-realization is to recognize nature's intrinsic value. Below, I explore these dimensions in detail.

To illustrate how human self-realization hinges on acknowledging the rights of other beings, Naess employs a thought experiment. He envisions a scenario in which a

[28] Susan Wolf, *Meaning in Life and Why It Matters* (Princeton University Press, 2010), pp. 61–62.

[29] Anthony Cross, *Ergo*, Vol. 8, no. 38, 2021, pp. 402–421.

mature human self, under appropriate conditions, willingly yields to a "more perfect" alien species. This hypothetical species lands on Earth, causing neither pain nor exploitation to other creatures, and shares essential traits with humanity. In such a case, Naess argues, humans could justifiably cede their dominant position to this superior form of life. He writes, "The ecosophical outlook is developed through an identification so deep that one's *own self* is no longer adequately delimited by the personal ego or the organism. One experiences oneself to be a genuine part of all life."[30] Naess further contends, "These propositions suggest that to ascribe value to animals, plants, landscapes, and wilderness areas independently of their relation to human utility or benefit is a philosophically legitimate procedure. To relate all value to mankind is a form of anthropocentrism which is not philosophically tenable."[31] These statements underscore his view that a fundamental human quality lies in recognizing other life forms, affirming their intrinsic value and right to self-realization. For Naess, a mature self emerges through this recognition, bound to others by a shared pursuit of fulfillment. Consequently, human self-realization depends on the thriving of all beings. In contrast, a false or constricted self—focused solely on personal or anthropocentric interests—neglects the welfare, inherent worth, and self-realization of other species. Naess' ecological self overcomes such limitations, cultivating a profound, interconnected sense of existence that embraces the broader web of life.

In this book, I examine the four principal types of self which are moral, cognitive, aesthetic, and ecological and I also explain the corresponding processes of self-transcendence as they have unfolded across the history of Western and Chinese philosophy. These conceptions constitute core elements of spirituality. Through a sustained historical investigation of these different formations of the self within distinct philosophical traditions, we can gain a deep understanding of spirituality as it appears in Emerson's thought and in Chinese philosophy. Although the moral self provides the primary framework for their respective understandings of selfhood, this moral dimension is never isolated. It remains deeply intertwined with cognitive, aesthetic, and ecological dimensions of the self. This interplay reveals that spiritual development in both traditions arises from the dynamic integration of these four aspects rather than from any single mode in isolation. In many other philosophers such as Spinoza and Descartes, the self tends to be one dimension.

4 Spiritual Practice

Spiritual practice can help people get rid of the false self and realize the true self, and finally get unity with ultimate values. In the history of Chinese and Western philosophy, many philosophers have roughly presented the following categories in

[30] Arne Naess, *Ecology, Community and Lifestyle: Outline of an Ecosophy*, trans. David Rothenberg (Cambridge University Press, 1989), p. 174.

[31] Ibid., p. 177.

the way of spiritual practices (1) rational spiritual practice; (2) trans-rational spiritual practice; (3) embodied and affective spiritual practice. The first refers to the cultivation of rationalism in spiritual practice, a typical example is Socratic dialogue. The second is trans-rationalist spiritual practice, which includes establishing a connection between the individual and God or the supernatural entity through practice, such as prayer. Body-emotionalist spiritual practice includes achieving a connection with the ultimate entity through meditation, sitting quietly, and rituals. Many spiritual practices in Chinese philosophy belong to this type. I will explain these spiritual practices separately below.

First, rational spiritual practice. The core feature of this spiritual practice is the emphasis on the training of rational critical thinking ability. Socratic dialogue is the most classic manifestation of this type. Here I take their main virtue in ancient Greece, courage, as an example to illustrate the main features of Socratic dialogue as spiritual practice. The general logic followed by the Socratic dialogue recorded by Plato is: (1) A person first proposes a definition of a concept, namely virtue; (2) Socrates then raises some counterexamples to question the rationality of this definition; (3) The person who proposes the definition further revises his not very perfect definition; (4) Socrates continues to question the new definition. In the end, the two sides still cannot reach a consensus on the exact definition of this concept, but they were one step closer to the truth. Socrates, while questioning the rationality of these definitions, is not as ignorant as he describes, but is guided by certain thoughts, which can be simply expressed as the ultimate good = virtue = knowledge = happiness. Speaking of the virtue of courage, their exploration of this definition follows the logic mentioned above.

1. Definition one: "If a man is willing to remain at his post and to defend himself against the enemy without running away, then you may rest assured that he is a man of courage."[32] Socrates questioned by using examples to illustrate that they were fighting against the enemy, but did not hold their positions and retreated. This is also courage. Courage is manifested in many aspects.
2. Definition two: "It would be wise endurance which could be courage."[33] Socrates questioned by the examples of smart spending of one's own money, doctors persistently refusing to let pneumonia patients and others eat and drink, to point out that this skillful persistence is not courage.
3. Definition three: "Courage is the knowledge not just of the fearful and the hopeful,…, it would be the knowledge of practically all goods and evils put together"[34] Socrates questioned: If courage is knowledge about the present and future good, then this knowledge cannot be obtained in a person who lacks temperance, justice, and piety.

---

32 Plato, *Laches and Charmides*, trans. Rosamond Kent Sprague (Hackett Publishing Company, 1992), p. 31.

33 Ibid., p. 35.

34 Ibid., p. 46.

The essence of Socratic dialogue is rationalist spiritual practice. When Socrates was in dialogue, he was not as he claimed, knowing his ignorance, but had metaphysical cognition in his mind. He believed that goodness is the ultimate existence of this world, transcending all existences. Specific moral points can become true morals only with goodness. Socratic dialogue has great demands on a person's virtue, which is restraint in dialogue. Restraint refers to maintaining humility when the values one has long adhered to are challenged. Without this humility, dialogue cannot be fundamentally guaranteed. Plato's dialogues record many people who left the dialogue in the middle because of anger or disagreement with their own values, while Socrates was able to maintain inner peace at all times. This peace is the embodiment of the virtue of people's pursuit of truth.

Second, trans-rational spiritual practice. This practice mainly refers to various ways of communication between humans and God or transcendental entities. This type of spiritual practice includes Christian prayer, devout confession, and Descartes-like introspection. Let's first look at the connection between man and God established by Augustine through prayer and confession. Augustine writes, "Every day, Lord, we are beset by these temptations. We are tempted without respite. The human tongue is our daily furnace. In this respect also you command us to be continent: grant what you command, and command what you will. In this matter you know the 'groaning' of my heart towards you, and the rivers which flow from my eyes."[35]

In Augustine's view, temperance is God's command, not the good chosen by individual rationality. Humans cannot defeat the pleasure brought by indulging desires with their own rational power, and thus cannot restrain desires. According to Augustine, it is impossible for humans to achieve happiness in this world. Only when life is in the expectation of the afterlife and the hopc of salvation is it happy. Augustine realizes his desire for self-restraint through prayer to God, not through his own rational restraint. Let's look at Augustine's confession again. "A delight to my eyes are beautiful and varied forms, glowing and pleasant colours. May these get no hold upon my soul; may God hold it! 'He has made these sights and they are very good."[36]

Augustine's confession is this kind of transcendental spiritual practice. For the spiritual practice of temperance, Augustine stresses that we should depend on God to enlighten our reason. We can pray to God to make our reason best guide our everyday life. Descartes' spiritual practice is also characterized by trans-rational style. His process of affirming human rationality is realized through transcendental introspection. His radical doubt is a manifestation of this introspective method. He affirms his existence through radical doubt, "But what then am I? A thing that thinks. What is that? A thing that doubts, understands, affirms, denies, is willing, is unwilling, and also imagines and has sensory perceptions."[37] Descartes' rational doubt is one type of spiritual practice which is transcendental. This is mainly reflected in his questioning inwardly what is the ultimate existence, and unquestionable. He does

---

[35] Saint Augustine, *Confessions*, trans. Henry Chadwick (Oxford University Press, 1991), p. 214.

[36] Ibid., pp. 209–211.

[37] René Descartes, *Meditations of First Philosophy*, trans. John Cottingham (Cambridge University Press, 1996), p. 19.

not adopt the way of rational individual dialogue, such as the veil of ignorance used by Rawls in the theory of justice. Instead he completely forces himself inwardly, and in his own internal dialogue process, he quoted the symbol of the devil's irrationalism. The spiritual practice of Emerson mentioned in this book belongs to this type; he emphasizes direct communication between man and God. He pointed out that there is a secret passage to God in everyone's heart, this secret passage refers to human intuition.

Third, embodied and affective spiritual practice. This type of spiritual practice refers to the establishment of communication between humans and the ultimate entity through the cultivation of human emotions and physical strength. Most Chinese philosophy belongs to this type. Let's first look at the spiritual practice of Confucianism, which belongs to the body-emotional type of practice. In these practices, the body needs to be present and is an important part of emotional cultivation. The rites of Confucian culture are this type of practice. First, it focuses on the presence of the body and the corresponding etiquette. "He did not sit, unless his mat was straight."[38] The emotional type of spiritual practice refers to the cultivation of love between people, this love is not the love of abstract rules, but the concrete love with the presence of the human body. One passage from *The Analects* states, "The Master said, 'Of neighbourhoods benevolence is the most beautiful.'"[39] When people live together in a place where moral people, that is, benevolent people live, then this is beautiful. It can be seen that Confucian spiritual practice has strong body-emotional characteristics. The spiritual practice of Daoism is similar to Confucianism, also focusing on the physical level. But it does not focus on the love that Confucianism continues to strengthen on the basis of rites; it pays more attention to natural love, such as the love of children for parents, and the care established on the basis of living together. Daoism also focuses on living together, we can see from the "world of virtue" described in Zhuangzi.

Spiritual practices in Emerson can be classified in trans-rational spiritual practice and embodied and affective spiritual practices. Emerson stresses solitude, expression and direct experience of nature, all of which tend to connect us to God. In Chinese philosophy, Confucius and Mencius stress embodied and affective spiritual practices through observing the rituals. Daoism stresses embodied spiritual practices through meditation and so on.

References

1. Thaddeus Metz, *Meaning in Life* (Oxford University Press, 2013).
2. Friedrich Wilhelm Nietzsche, *The Will to Power*, ed. Walter Kaufmann and R. J. Hollingdale (Vintage Books, 1968).
3. Suan Wolf, *Meaning in Life and Why It Matters* (Princeton University Press, 2010).

[38] Confucius, *The Analects*, trans. D. C. Lau (Penguin Books, 1979), p. 103.
[39] Ibid., p. 72.

4. Krzysztof Piotr Skowroński, *Values and Powers: Re-reading the Philosophical Traditional of American Pragmatism*, Amsterdam-New York, NY 2009).
5. Joe Mintoff, "Transcending Absurdity", *Ratio* XXI 1.
6. Immanuel Kant, *Groundwork of the Metaphysics of Morals* (Cambridge University Press, 2011).
7. Tom Regan, *The Case for Animal Rights* (University of California Press, 1983).
8. Holmes Rolston III, *Environmental Ethics: Duties to and Values in the Natural World* (Temple University Press, 1988).
9. Eugene C. Hargrove, *Foundations of Environmental Ethics* (Prentice Hall, 1989).
10. Victoria Jenkins, In Defense of Natural Beauty: Aesthetic Obligation and the Law on the Designation of Protected Landscapes in England and Wales, *Environmental Law Review*, Volume 11, Issue 1, 2020.
11. Plato, *Republic*, trans. G. M. A. Grube (Hackett Publishing, 1992).
12. John Dewey, *Experience and Nature* (W. W. Norton & Company, 1929).
13. Zhuangzi, *The Complete Works of Zhuangzi*, trans. Burton Watson (Columbia University Press, 1968).
14. Guyton B. Hammond, *Power of Self-transcendence: An Introduction to the Philosophical Theology of Paul Tillich* (The Bethany Press, 1966).
15. Jennifer A. Frey and Candace Vogler (Eds), *Self-transcendence and Virtue: Perspectives from Philosophy, Psychology, and Theology* (Routledge, 2018).
16. Arne Naess, *Ecology, Community and Lifestyle: Outline of an Ecosophy, trans. David Rothenberg* (Cambridge University Press, 1989).
17. Viktor E. Frankl, "Self-transcendence As A Human Phenomenon", *Journal of Humanistic Psychology*, 1966, 6 (2).
18. Confucius, *The Analects*, trans. D. C. Lau (Penguin Books , 1979).
19. *A Source Book in Chinese Philosophy*, trans. Wing-tsit Chan (Princeton University Press, 1963), p. 104.
20. Plato, *Laches and Charmides*, trans. Rosamond Kent Sprague (Hackett Publishing Company, 1992).
21. Saint Augustine, *Confessions*, trans. Henry Chadwick (Oxford University Press, 1991).
22. René Descartes, *Meditations of First Philosophy*, trans. John Cottingham (Cambridge University Press, 1996).

Chapter 3
Emerson: American Transcendentalism and Aesthetic Revolution

Emerson, widely regarded as the founder of American Transcendentalism, initiated an aesthetic revolution through transforming the traditional negative attitudes towards wilderness into wilderness appreciation. However, Emerson occupied an ambiguous position in philosophical circles for much of his legacy. Unlike traditional Western philosophers, Emerson eschewed systematic analysis, logical argumentation, and conceptual synthesis. Instead, his writing is characterized by a fragmented style, rich with interdisciplinary references drawn from literature, religion, and science. Rather than rigorously examining the terms he borrowed, he employed them selectively to support his insights, a method that diverged sharply from conventional philosophical rigor. Consequently, many American university philosophy departments—including Harvard's, despite naming its philosophy building "Emerson Hall" in 1905 and housing a statue of him crafted by Frank Duveneck—have hesitated to classify him as a philosopher in the strict sense.

This perception stems partly from assessments like that of philosopher Stanley Cavell, who argued that Emerson's work lacks significant contributions to core philosophical domains such as metaphysics, epistemology, ethics, philosophy of science, or practical philosophy. Instead, Emerson is often cast as a luminous transcendentalist, a towering public intellectual, and a prose stylist whose poetic language captivates readers. Yet, sustained scholarly attention from figures like Stanley Cavell and Cornel West has begun to reshape this view, highlighting his broader cultural and intellectual significance.

Cornel West offers a nuanced appraisal: "Emerson lacks the patience and persistence to be a great poet. He does not have the deep sense of alienation and marginality to be a profound prophet. And he does not possess the talent for logical precision and sustained argumentation to be a rigorous philosopher. Yet Emerson is more than a mediocre man of letters or a meteoric man of lectures. Rather he is a cultural critic who devised and deployed a vast array of rhetorical strategies in order to exert intellectual and moral leadership over a significant segment of the educated classes of his

S. Gao, *Comparative Studies Between Emerson and Chinese Philosophy*, Nature Aesthetics: Interdisciplinary and Intercultural Dialogue,
https://doi.org/10.1007/978-981-95-0421-3_3

day."[1] West's analysis positions Emerson as a transformative figure—not a philosopher in the traditional mold, but a cultural architect whose rhetorical ingenuity forged a new American identity.

Although Emerson's status as a philosopher remains controversial, I argue that the transcendentalist philosophy he founded plays a pivotal role in advancing a revolution in nature aesthetics. While he doesn't construct a systematic theory of nature aesthetics, Emerson integrates multidisciplinary perspectives to describe the everyday aesthetic experience of nature, thereby effecting a revolution in aesthetic appreciation of nature, especially wilderness appreciation. This revolution has direct influence on the establishment of the American Wilderness Act of 1964. In the following I will examine Emerson's thought from two aspects: (1) Emerson and American Transcendentalism; (2)Emerson's Aesthetic Revolution.

1 Emerson and American Transcendentalism

The word "transcendentalism" was not named by Emerson himself, nor by Emerson's peers, but by a monotheistic pastor who opposed transcendentalist philosophy. He used the concept of transcendentalism with a sense of irony. This person is Francis Bowen. He commented on Emerson's "On Nature" in "Christian Examiner". He writes, "The aim of the Transcendentalists is high. They profess to look not only beyond facts, but without the aid of facts, to principles."[2] Transcendentalism is mainly expressed in two ways: One is in the form of a transcendentalist club, which is a club organized by Emerson from 1836 to 1840. This club is not an institutionalized existence, it did not flourish in universities or seminaries, but was held in turn in the homes of transcendentalist members. The other is the journal of *Dial*. In the following, I will explore American transcendentalism from the following three aspects: (1) the origins of American transcendentalism; (2) opposition to the traditional religions; (3) the concepts of American transcendentalism.

First, the philosophical idea of American transcendentalism is derived from Immanuel Kant. In his essay "The Transcendentalist", delivered as a lecture that year and published in 1842, Emerson credits Kant as the inspiration for transcendentalism's idealist orientation. He writes, "It is well known to most of my audience, that the Idealism of the present day acquired the name of Transcendental, from the use of that term by Immanuel Kant, of Konigsberg, who replied to the skeptical philosophy of Locke, which insisted that there was nothing in the intellect which was not previously in the experience of the senses, by showing that there was a very important class of ideas, or imperative forms, which did not come by experience, but through which experience was acquired; that these were intuitions of the mind itself;

[1] Cornel West, *The American Evasion of Philosophy: American Genealogy of Pragmatism* (Madison: University of Wisconsin Press, 1989), p. 10.

[2] *The Recognition of Ralph Waldo Emerson: Selected Criticism Since 1837*, ed. Milton R. Konvitz (Ann Arbor: The University of Michigan Press, 1972), p. 4.

and he denominated them the *Transcendental* forms." [3] Here, Emerson underscores the mind's innate capacities, aligning with Kant's rebuttal of Lockean empiricism by emphasizing a priori intuition that structures experience.

Emerson differs significantly from Kant in his conception of reason's source and ultimate aim. For Emerson, reason as the mind's highest faculty originates not within itself but in the Oversoul, a universal spiritual unity. He writes, "The soul is the perceiver and revealer of truth."[4] This capacity manifests only through self-denial, enabling recognition of the ultimate ontology—God. In contrast, Kant rejects such an ultimate ontology as knowable. For Kant, the "thing-in-itself" represents the limit of cognition: "We can never have knowledge of things as they are in themselves, independent of our perception."[5] While Kant invokes God, this deity is a rational postulate, not a preexistent reality: "The idea of God arises as a necessary condition of practical reason, not as an object of theoretical knowledge."[6] In Kant's system, human reason holds absolute dignity, autonomous and prior to any external divine entity, whereas Emerson subordinates reason to the Oversoul's transcendent presence.

Second, Emerson's transcendentalism emerged in the background of opposing the dominant religious paradigms of his era, notably historical Christianity including Calvinism and Unitarian monotheism. Calvinism, rooted in the doctrine of divine sovereignty, asserts God's absolute dominion over humanity, demanding unwavering obedience to His will and reason as the arbiter of moral action. Emerson recoiled at its stifling effect on human creativity, critiquing its narrow worldview: "Our religion vulgarly stands on numbers of believers. Whenever the appeal is made,— no matter how indirectly,—to numbers, proclamation is then and there made that religion is not. He that finds God a sweet enveloping thought to him never counts his company. When I sit in that presence, who shall dare to come in? When I rest in perfect humility, when I burn with pure love, what can Calvin or Swedenborg say?"[7] For Calvinism, human nature is originally sinful and God chooses who will be saved. The salvation of people relies on Church doctrines and God. However, for Emerson, each person has the ability to have direct communication with God without relying on authority.

Emerson also challenges Unitarian monotheism, a liberal offshoot of Calvinism that emerged in the late eighteenth century, replacing the Trinitarian doctrine of Father, Son, and Holy Spirit with a singular, rational God. While Unitarianism affirms human reason and dignity—offering a humanistic foundation—Emerson found its emphasis on dry rationalism wanting. He states in the essay "The Divinity School Address", "Historical Christianity has fallen into the error that corrupts all attempts to communicate religion. As it appears to us, and as it has appeared for ages, it is not

[3] Ralph Waldo Emerson, *Essays and Lectures*, ed. Joel Porte (The Library of America, 1983), p. 198.

[4] Ralph Waldo Emerson, *The Portable Emerson*, ed. Jeffrey S. Cramer (Penguin Books, 2014), p. 217.

[5] Kant, *Critique of Pure Reason*, trans. Norman Kemp Smith (St. Martin's Press, 1965), p. 82.

[6] Kant, *Critique of Practical Reason, trans.* Lewis White Beck (Bobbs-Merrill 1956), 129 [5:124]

[7] Ralph Waldo Emerson, *The Portable Emerson*, ed. Jeffrey S. Cramer (Penguin Books, 2014), p. 224.

the doctrine of the soul, but an exaggeration of the personal, the positive, the ritual. It has dwelt, it dwells, with noxious exaggeration about the *person* of Jesus."[8] For Emerson, each person has a soul, it will be able to perceive the truth, goodness, and beauty directly without relying on interpretations from rationality and rituals.

Central to Emerson's critique is his reimagining of divine revelation. Unlike Calvinism's reliance on miracles or Unitarianism's rational proofs, he locates God's presence in moral sentiment of human beings, "A more secret, sweet, and overpowering beauty appears to man when his heart and mind open to the sentiment of virtue. Then he is instructed in what is above him."[9] For Emerson, the presence of spiritual laws, which is also termed as God will inspire a moral sentiment among human beings. When human mind is open to the moral sentiment, human beings can feel the overpowering beauty. This intuitive communion elevates the individual, as he asserts, "The simplest person who in his integrity worships God, becomes God; yet for ever and ever the influx of this better and universal self is new and unsearchable. It inspires awe and astonishment."[10] Here, self-elevation through spiritual authenticity aligns human status with the divine, contingent not on external authority but on inner transformation. From this critique of traditional religious values, Emerson's transcendentalism constructs a novel framework encapsulated in a "new Trinity" of man, nature, and God. Unlike Calvinism's anthropomorphic deity, Emerson's God is a living consciousness, manifest as ultimate values—metaphysically encompassing truth, goodness, and beauty. He writes, "The Supreme Critic on the errors of the past and the present, and the only prophet of that which must be, is that great nature in which we rest as the earth lies in the soft arms of the atmosphere; that Unity, that Over-Soul, within which every man's particular being is contained and made one with all other;"[11] This pantheistic vision repositions nature as the medium through which humans apprehend God, supplanting Unitarianism's rationalist approach with experiential insight.

Third, in terms of the definition of American transcendentalism, I would like to mention Shoei Ando's viewpoint. He defines it as "a religious philosophy with a concept of God, asserting the primacy of the spiritual over the material, but rejecting the concept of trinity, that is to say, the concept of the union of God the Father, Jesus Christ the Son, and the Holy Ghost."[12] "It affirms that man can be directly connected with God at the bottom, or in the recesses, of the human heart, without the help of mediator like Jesus Christ,..."[13] Shoei Ando's definition misses one important element of nature. It is true that Emerson rejects the idea of trinity as interpreted

---

[8] Ralph Waldo Emerson, *Essays and Lectures*, ed. Joel Porte (The Library of America, 1983), pp. 80–81.

[9] Ibid., pp. 75–76

[10] Ralph Waldo Emerson, *The Portable Emerson*, ed. Jeffrey S. Cramer (Penguin Books, 2014), p. 223.

[11] Ibid., p. 211.

[12] Shoei Ando, *Zen and American Transcendentalism: An Investigation of One's Self* (The Hokuseido Press, 1970), p. 186.

[13] Ibid.

by traditional Christianity but he affirms the new trinity among humanity, nature, and God. Nature becomes the new mediator between humanity and God. Emerson writes: "Since every thing in nature answers to a moral power, if any phenomenon remains brute and dark it is because the corresponding faculty in the observer is not yet active."[14] For Emerson, human beings can know God through interpretation of nature. Scientific inquiry aids this interpretation, yet he cautions against its limitation: "Empirical science is apt to cloud the sight, and by the very knowledge of functions and processes to bereave the student of the manly contemplation of the whole. The savant becomes unpoetic."[15] Poets, he argues, can interpret this text, blending rational precision with imaginative depth. "The poet is the sayer, the namer, and represents beauty. He is a sovereign, and stands on the centre."[16] For Emerson, poets are representative of ideal persons. He can use both scientific knowledge and poetic imagination to convey the spirit of nature by words.

2 Emerson's Aesthetic Revolution

Emerson promotes aesthetic revolution in his lifetime. Wilderness becomes the object of aesthetic appreciation and a meaningful place. In the following, I will examine this point in two aspects: (1) traditional attitude towards wilderness appreciation and Emerson's aesthetic revolution; (2) historical background of Emerson's aesthetic revolution.

First, in Emerson's lifetime, people's attitude towards wilderness is deeply influenced by religious views especially puritanism and the hardship of life in the frontier. Peter N. Carroll in his book titled *Puritanism and the Wilderness: The Intellectual Significance of the New England Frontier 1629*–1700 gives a systematic overview of this topic. According to Peter, wilderness is a place of evil forces and symbol of sin for Puritans, "The Puritan colonists translated all adversity as signs of this diabolic opposition and conceived of the wilderness as the site of moral battle against Satan and his minions."[17] The extreme hardship of wilderness is also another impediment for people to appreciate the beauty of wilderness. Emerson promotes aesthetic revolution in appreciation of wilderness by his metaphysical framework.

From the discussion of chapter one, we can see that in the history of Western philosophy, aesthetic appreciation of physical nature has been neglected for long time since the metaphysics of traditional Western philosophy tends to favor the intelligible world rather than the physical world. These traditions can be traced back to Plato's metaphysics and neglect of the philosophical importance of material world.

---

[14] Ralph Waldo Emerson, *The Portable Emerson*, ed. Jeffrey S. Cramer (Penguin Books, 2014), p. 243.

[15] Ibid., pp. 48-49.

[16] Ibid., pp. 239.

[17] Peter N. Carroll, *Puritanism and The Wilderness: The Intellectual Significance of the New England Frontier, 1629–1700* (New York & London: Columbia University Press,1969), p. 73.

In the following, I will mention Plato's metaphysics first since it is the origin of metaphysics of traditional Western philosophy. Although many philosophers have diverged from Plato's metaphysics, the basic metaphysical spirit is the same, that is, dualism. Intelligible world is mostly valued than the physical world. The basic idea of Platonic metaphysics is that true existence lies in an ideal realm of Forms and the material nature is just imitation of the forms.

Plato's neglect of natural aesthetics can be found in *Phaedrus*. The dialogue opens with Socrates and Phaedrus meeting outside Athens, where Phaedrus recounts a speech on love by Lysias. Intrigued, Socrates agrees to discuss it in a tranquil suburban setting. Upon reaching a shaded spot, Socrates remarks, "By Hera, it is a charming resting place. For this plane tree is very spreading and lofty, and the tall and shady willow is very beautiful, and it is in full bloom, so as to make the place most fragrant; then, too, the spring is very pretty as it flows under the plane tree, and its water is very cool, to judge by my foot. And it seems to be a sacred place of some nymphs and of Achelous, judging by the figurines and statues. Then again, if you please, how lovely and perfectly charming the breeziness of the place is!"[18] Phaedrus responds, "You are an amazing and most remarkable person. For you really do seem exactly like a stranger who is being guided about, and not like a native. You don't go away from the city out over the border, and it seems to me you don't go outside the walls at all."[19] Socrates replies, "Forgive me, my dear friend. You see, I am found of learning. Now the country places and the trees won't teach me anything, and the people in the city do."[20] This exchange reveals Socrates' indifference to nature's beauty; his fascination lies not in the landscape but in the rational pursuit of ideas, epitomized by the discourse Phaedrus offers.

Plato reinforces this stance in *Phaedo*, contrasting the flawed material world with the perfection of the forms. He writes, "But the fact is the same, that by reason of feebleness and sluggishness, we are unable to attain to the upper surface of the air; for if anyone should come to the top of the air or should get wings and fly up, he could lift his head above it and see, as fishes lift their heads out of the water and see the things in our world, so he would see things in that upper world; and, if his nature were strong enough to bear the sight, he would recognize that that is the real heaven and the real light and the real earth."[21] In Plato's cosmology, ideal beauty resides in the Forms, as exemplified by trees of flawless proportions and smooth and unbroken surfaces, whereas material nature merely shadows this perfection and lacks intrinsic aesthetic worth. This geometric idealism, favoring mathematical harmony over nature's irregularity, clashes with the organic beauty celebrated by later aestheticians.

Emerson, though indebted to Plato's idealism, radically departs from this view, ascribing beauty to all natural forms irrespective of mathematical precision. He

[18] Plato, *Plato In Twelve Volumes (I), Euthyphro Apology Crito Phaedo Phaedrus*, trans. Harold North Flower (Harvard University Press, 1914), p. 423.

[19] Ibid., p. 423.

[20] Ibid., pp. 423–425

[21] Ibid., p. 377.

asserts, "Every natural fact is a symbol of some spiritual fact."[22] Emerson regards material nature not as illusory as Plato interprets, but as possessing genuine metaphysical reality. It is the result from God (the Over-Soul).[23] This ontological affirmation provides him the possibility to fully appreciate the beauty of the physical world. However, there is another element which makes him appreciate the beauty of material nature. For him, God, is not value-neutral but embodies the ultimate values of truth, goodness, and beauty. For Emerson, the aesthetic appreciation of nature is a vital pathway to understanding divine goodness.

The aesthetic revolution sparked by transcendentalist philosophy played a pivotal role in fostering the protection of wilderness in America. Unlike traditional philosophical schools, transcendentalism does not construct a comprehensive metaphysical system. Nevertheless, its metaphysical perspective on nature can be discerned through the concepts it employs to describe the natural world. Moreover, Transcendentalism's view of human nature is deeply intertwined with its understanding of nature, positing a return to nature as life's ultimate purpose. This dual emphasis on nature and humanity directly fueled America's aesthetic appreciation and legislative safeguarding of the wilderness.

Second, to fully grasp Emerson's aesthetic revolution, it is essential to contextualize the era in which he lived. A historian contemporary with Emerson described the prevailing American ethos of the time as centered on an ideal of the "dream"—a collective aspiration deemed destined for realization. This dream galvanized intellectuals and the broader public alike. He writes, "And although the particulars of each American's dream certainly varied from one person to the next, Americans collectively dreamed of a 'great democracy' that would 'transmute its social power into the higher forms of thought,' 'provide for the moral and intellectual needs of mankind,' 'take permanent political shape,' 'give new life to religion and art,' and produce 'a higher variety of the human race.'"[24] This vision, framed as the ultimate aim of a "great democracy," inspired diverse expressions within the transcendentalist movement Emerson spearheaded.

Within the transcendental club, different figures pursued their dream in distinct ways. Some, like George Ripley and Amos Bronson Alcott, sought to realize it through politically and religiously infused social reforms, exemplified by the utopian communities of Brook Farm and Fruitlands. George Ripley, a former Unitarian minister and transcendentalist, envisioned Brook Farm as a transcendental utopia—an agricultural and educational collective where harmonious coexistence was paramount. In a letter to Emerson, he articulated its aims: "Our objects, as you know, are to ensure a more natural union between intellectual and manual labor than now exists; to combine the thinker and the worker, as far as possible, in the same

---

22 Ralph Waldo Emerson, *The Portable Emerson*, ed. Jeffrey S. Cramer (Penguin Books, 2014), p. 28.

23 Spinoza, *A Spinoza Reader: The Ethics and Other Works*, trans. Edwin Curley (Princeton University Press,1994), p. 94.

24 Joel Myerson, *A Historical Guide to Ralph Waldo Emerson* (Oxford University Press, 2000), p. 47.

individual; to guarantee the highest mental freedom, by providing all with labor, adapted to their tastes and talents, and securing to them the fruits of their industry; to do away with the necessity of menial services, by opening the benefits of education and the profits of labor to all; and thus to prepare a society of liberal, intelligent, and cultivated persons, whose relations with each other would permit a more simple and wholesome life, than can be led amidst the pressure of our competitive institutions."[25] Ripley's vision emphasized integrating intellectual and physical labor, ensuring equitable access to education and work, and fostering a community free from the strains of competitive society. Though Emerson admired such initiatives, he wavered and ultimately declined to join Brook Farm.

Emerson showed greater sympathy toward Fruitlands, another utopian experiment led by Alcott. This commune prioritized self-reliance, simplicity, and industriousness, aiming to avert moral corruption and create a peaceful, virtuous community. Despite its religious undertones, Emerson hesitated for days before opting not to participate. While he engaged in political efforts like the abolition of slavery, his primary focus remained the advancement of thought itself rather than direct involvement in communal experiences. Emerson's devotion to thought and writing directly influences his student Thoreau, who put his ideas of American transcendentalism into practice through living in Walden pond and everyday aesthetic observation of wilderness.

References

1. Cornel West, *The American Evasion of Philosophy: American Genealogy of Pragmatism* (Madison: University of Wisconsin Press, 1989).
2. *The Recognition of Ralph Waldo Emerson: Selected Criticism Since 1837*, ed. Milton R. Konvitz (Ann Arbor: The University of Michigan Press, 1972).
3. Ralph Waldo Emerson, *Essays and Lectures*, ed. Joel Porte (The Literary Classics of America, 1983).
4. Ralph Waldo Emerson, *The Portable Emerson*, ed. Jeffrey S. Cramer (Penguin Books, 2014).
5. Kant, *Critique of Pure Reason*, trans. Norman Kemp Smith (St. Martin's Press, 1965).
6. Kant, *Critique of Practical Reason*, trans. Lewis White Beck (Bobbs-Merrill, 1956).
7. Emerson, *The Portable Emerson*, ed. Jeffrey S. Cramer (Penguin Books, 2014).
9. Peter N. Carroll, *Puritanism and The Wilderness: The Intellectual Significance of the New England Frontier*, 1629–1700 (New York & London: Columbia University Press,1969).
10. Plato, *Plato In Twelve Volumes (I), Euthyphro Apology Crito Phaedo Phaedrus*, trans. Harold North Flower (Harvard University Press, 1914).
11. Joel Myerson, *A Historical Guide to Ralph Waldo Emerson* (Oxford University Press, 2000).
12. Octavius Books Frothingham, *George Ripley* (Houghton Mifflin Company, 1882).

[25] Octavius Books Frothingham, *George Ripley* (Houghton Mifflin Company, 1882), p. 307.

Chapter 4
Foundations for a Comparative Study Between Emerson and Chinese Philosophy

Over the past several decades, there has been a growing scholarly interest in conducting comparative studies between Emerson and Chinese philosophy. Many works have been published in this field such as Arthur Versluis' *American Gurus: From Transcendentalism and New Age Religion* (2014), Mansu Qian's *Emerson and China: Reflections on Individualism* (2018), and Yoshio Takanashi's *Emerson and Neo-Confucianism* (2014). In this book, I undertake a comparative examination of these philosophical traditions through the conceptual lens of nature, spirituality, and place. This approach raises an important methodological question: what are the philosophical foundations that justify comparative studies between Emerson and Chinese philosophy? I argue that there are three foundations. First, it enables us to gain a deep understanding of the conception of nature and the role it plays in the construction of self in Emerson and Chinese philosophy; Second, it offers valuable insights that can facilitate self-transcendence and contribute to the reconstruction of the modern self; Third, it encourages us to reflect critically on our own aesthetic appreciation of nature and the role it plays in self-construction. It can help us develop a new self through intercultural dialogue.

1 Nature and Construction of Self

In conducting this comparative study between Emerson and Chinese philosophy, I begin with the concept of nature since it constitutes the foundation for spirituality in both Emerson's philosophy and Chinese thought. Their pursuit of spirituality can be summarized in one statement: Virtue represents the idea of self, which is attained through living in harmony with nature. Both Emerson and Chinese philosophy regard a life of living in harmony with nature as the essence of a meaningful life. Emerson explicitly addresses this in his essay "The Conduct of Life," where he situates power at the core of life's meaning: "Life is a search after power; and this is an element

S. Gao, *Comparative Studies Between Emerson and Chinese Philosophy*, Nature Aesthetics: Interdisciplinary and Intercultural Dialogue,
https://doi.org/10.1007/978-981-95-0421-3_4

with which the world is so saturated,—there is no chink or crevice in which it is not lodged,—that no honest seeking goes unrewarded."[1] This pursuit of power, a recurring theme in Emerson's self-reflection, is further illuminated in Richardson's biographical account: "Emerson's lifelong search, what he called his heart's inquiry, was 'Whence is your power?' His reply was always the same: 'From my nonconformity. I never listened to your people's law, or to what they call their gospel, and wasted my time. I was content with the simple rural poverty of my own. Hence this sweetness.'"[2] For Emerson, power emerges from rejecting conventional values and forging a new moral framework, rooted in natural aesthetics.

Emerson's philosophy focuses on the interplay between humanity and nature. In the essay "The Method of Nature", he writes, "The festival of the intellect and the return to its source cast a strong light on the always interesting topics of Man and Nature."[3] He views material nature as secondary to the spiritual law, which holds metaphysical primacy: "In the absence of man, we turn to nature, which stands next. In the divine order, intellect is primary; nature, secondary; it is the memory of the mind. That which once existed in intellect as pure law, has now taken body as Nature."[4] Nature, for Emerson, mirrors the human mind, serving as a medium to reconnect with God. This idea resonates with Plato's theory of recollection where knowledge is recalled from a prior state obscured by bodily existence.

In contrast, Chinese philosophy, through the "unity of Heaven and man," also prioritizes harmony with nature but interprets it differently. Wing-tsit Chan asserts, "If ONE WORD could characterize the entire history of Chinese philosophy, that word would be humanism—not the humanism that denies or slights a Supreme Power, but one that professes the unity of man and Heaven."[5] Historian Yingshi Yu reinforces this, citing Qian Mu's late life epiphany, "In September 1989, Mr. Qian suddenly had his last 'epiphany' in his life, affirming that the view of 'the unity of heaven and man' is 'the destination of the entire Chinese traditional cultural thought.' He was very excited about this 'epiphany,' and finally completed a spoken short essay three days before the Dragon Boat Festival in 1990, titled 'The Possible Contribution of Chinese Culture to the Future of Mankind'".[6] Yu traces this concept's evolution, noting its profound influence on Chinese selfhood and aesthetics.

Despite their shared focus on harmony, Emerson and Chinese philosophy diverge fundamentally. For Emerson, understanding nature is not merely a means to achieve harmony, but an essential pathway to glimpsing the divine. He writes, "We may therefore safely study the mind in nature, because we cannot steadily gaze on it

---

[1] Ralph Waldo Emerson: *Essays and Lectures*, ed. Joel Porte (The Library of America, 1983), p. 971.

[2] Robert D. Richardson Jr., *Emerson: The Mind on Fire* (University of California Press, 1995), p. 4.

[3] Ralph Waldo Emerson, *The Portable Emerson*, ed. Jeffrey S. Cramer (Penguin Books, 2014), p. 97.

[4] Ibid., p. 98.

[5] Wing-tsit Chan, *A Source Book in Chinese Philosophy,* trans. Wing-tsit Chan (Princeton University Press, 1963), p. 3.

[6] Yingshi Yu, *Between Heaven and Man: An Inquiry into the Origins of Ancient Chinese Thought* (《论天人之际:中国古代思想起源试探》) (Zhonghua Book Company, 2014).

in mind; as we explore the face of the sun in a pool, when our eyes cannot brook his direct splendors."[7] This recalls Plato's Allegory of the Cave, where the sun—symbolizing ultimate goodness—is approached indirectly through reflections. For Emerson, nature's beauty, informed by scientific insight and intuition, bridges the human mind to the divine. Conversely, in Confucianism and Daoism, harmony stems not from nature's material essence but from the values of Heaven and Earth, realized through internal transformation rather than external cognition. In Daoism, nature reflects the ceaseless flux of *qi*, governed by the Dao, not an objective law, while Confucianism grounds understanding in practical experience rather than aesthetic contemplation. Emerson ties nature to the moral self, asserting, "…therefore, that spirit, that is, the Supreme Being, does not build up nature around us, but puts it forth through us, as the life of the tree puts forth new branches and leaves through the pores of the old."[8] Here, nature and the mind share a divine origin, with nature acting as a mediator for recollection. This is distinct from the Chinese philosophical perspective, in which material nature does not function as a direct and dualistic counterpart to the self. Through comparative study, we can see that the concept of nature is culturally constructed. Their interpretation of nature is deeply influenced by their metaphysics. Emerson's interpretation of nature is derived from the concept of traditional Christian God. He absorbs many scientific elements to give a new interpretation of nature. Nature becomes the new interpreter of God.

In Chinese philosophy, the metaphysical concept of Dao influences the interpretation of nature. Their interpretation of nature is not just for the sake of interpretation, rather it is to make sense of human world. The final purpose is to live a good life. Through comparative studies of nature, we can gain a deep understanding of nature and spirituality in each culture. This understanding is the precondition to generate a new interpretation of nature. Comparative studies from the concept of nature can also help us better reflect on the modern self which has lost connection to nature. In terms of the modern self, Louis A. Sass remarked on Charles Taylor's two modern conceptions of self, autonomous and the expressivist image of man. He points out that "Thus, each of these images entails a conception of the human subject as essentially *self*-defining rather than as defined by its relationship to some all-encompassing comic or religious order."[9] In a era of rapid technological advancement and urbanization, nature has come to play a diminishing role in the process of self-definition. Individuals increasingly construct their conceptions of self and the good life independently of the natural world. When nature is severed from many people's sense of self-identity and their understanding of the good life, environmental crisis tends to intensify.

---

[7] Ralph Waldo Emerson, *The Portable Emerson*, ed. Jeffrey S. Cramer (Penguin Books, 2014), p. 98.

[8] Ibid., p. 47.

[9] *Constructions of the Self*, ed. George Levine (New Brunswick, New Jersey: Rutgers University Press, 1992), p. 20.

2 Self-transcendence and Reconstruction of Self

Comparative studies between Emerson and Chinese philosophy can promote self-transcendence and reconstruction of the self. This point is related to the following questions I am going to examine here: (1) How is the self shaped by culture? (2) How can we understand deeply the cultural construction of the self through self-reflection and what is the final destination of self-reflection? (3) How can we reconstruct the self?

First, many scholars have generally agreed that the self is constructed by culture. In terms of the culture which has shaped the essence of the self, it contains many elements including politics, architecture, economics, language, and geographical environment and so on. All of these elements are a reflection of core values, which is deeply influenced by philosophy, especially metaphysics. Values determine the ways of interaction among the people, design of architecture, landscape, politics and so on. Through comparative studies between Emerson and Chinese philosophy, we can gain a deep understanding of the importance of metaphysics and values which have shaped the essence of the self. For example, Emerson's concept of self-reliance is deeply influenced by his interpretation of God. His idea of God is no longer traditional Christian God, rather it is the life of consciousness of ultimate truth, goodness, and beauty. These values have profound influence on his concept of self. In Chinese philosophy, the metaphysical concept of Heaven and Earth is the moral source of the concept of self.

Second, we can gain a deep understanding of self through self-reflection. Self-reflection needs a conceptual tool which can be provided by philosophy. In this book, I will use the conceptual analysis to examine the concept of self in Emerson and Chinese philosophy. I will focus on the core concepts that are related to understanding of the self which include nature, power, value, virtue, self-transcendence, and place. However, I would like to stress that self-reflection is not the final destination. We will not think for the sake of thinking alone. Instead, we think in order to better conduct our everyday life.

Third, our deep understanding of the self can promote self-transcendence and construct a new self. The construction of new self needs three elements: (1) metaphysical foundation. Metaphysics is the first step for us to achieve self-transcendence and undertake the great project of self-reconstruction. The comparative study between Emerson and Zhuangzi can provide an insight for us to have a new metaphysics. The new metaphysics need to integrate science and values; (2) I argue that both Emerson and Chinese philosophy regard virtue as a core element of the self and place the pursuit of virtue at the center of spiritual life. Through a comparative study, this book proposes the concept of an eco-aesthetic self, which represents a transcendence of the traditional virtue-based self which is found in both Emerson and Chinese philosophy. This eco-aesthetic self encourages individuals to appreciate the intrinsic value of nature itself and to integrate this appreciation into their spirituality. In other words, eco-aesthetic self can generate a new type of virtue: environmental virtue; (3) Spiritual Practices. Without the spiritual practices, the new self cannot take root. I stress the ecological aesthetic practices for the new self which will motivate people to develop environmental virtues which are the new direction for developing the self.

3 Reflection of Nature Aesthetics and Its Influence on the Concept of Place

The comparative studies can help us have a reflection of nature aesthetics in different cultures. In each culture, people have a unique way of aesthetic appreciation of nature, which is deeply shaped by their philosophy and religion. Aesthetic appreciation of nature has a direct influence on the concept of place. When we gain a deep understanding of it, it can help us better understand how self is shaped by aesthetic appreciation and how this self is connected to place. This exploration can enrich our understanding of the concept of place. With this understanding, we can create a new place and also try to conserve the old place, which has equal value. Emerson initiates the aesthetic revolution and wilderness becomes the object of aesthetic appreciation. It is no longer treated as geological space rather than a meaningful place. While in Chinese philosophy, aesthetic appreciation is usually directed to cultivated place such as farming land and gardens.

This study culminates in the concept of place, aiming to integrate theory into practice. From part I of this book, we can see that the interpretation of nature by many Western philosophers tends to neglect the concept of place and its role in construction of self-identity. They tend to prioritize reason in defining the self. However, many philosophers today have challenged this insight. For example, Edward S. Casey asserts, "In the past three centuries in the West–the period of 'Modernity'–place has come to be not only neglected but actively suppressed. Owing to the triumph of the natural and social sciences in this same period, any serious talk of place has been regarded as regressive or trivial."[10] Edward Casey points out that many people have neglected the concept of place due to the development of modernity and modern science. He advocates that we should put the place as the very center of philosophical exploration. In China, although Confucianism and Daoism's interpretations of Heaven and Earth are closely related to the concept of place and its role in self-identity, the rapid development of urbanization and virtual realities has increasingly posed a big challenge to this identity. The creation of national park system in China also opens vast wilderness place to people which will bring the new possibility for construction of a new ecological self.

Although conservation of traditional farmland, villages, and wilderness has become national polices, the philosophical foundations for conservation of these places are not well developed. Many philosophical questions remain to be answered and explored deeply to justify the conservation of these places. In terms of well-cultivated nature such as farmland and gardens, philosophical questions need to be answered if we want to have an effective conservation of these lands: should we

[10] Edward Casey, *Getting Back into Place* (Bloomington: Indiana University Press, 1993), p. xiv.

spend more money in conservation of them? What are justifications for establishing wilderness laws in China? What are the wilderness areas? Can wilderness become a new type of place for Chinese culture? Holmes Rolston III values wilderness more and he argues, "On these occasions natural things are not used, at least not used up, to satisfy human needs. Rather, they are valued, when humans encounter them, for what they are in themselves, and not just for the sake of human appreciation."[11] However, in China, traditional Chinese philosophy promotes valuing cultivated nature while neglects the aesthetic appreciation of wilderness. Through the intercultural dialogue, we can have self-reflection of aesthetic appreciation and transcend the traditional ways of appreciation and construct a new sense of place while not neglecting the old sense of place.

Right now in China, people become more attached to cultivated places such as cities, farmland, and gardens. Professor Yuedi Liu has advocated everyday aesthetics in his research. Everyday aesthetics is closely related to appreciation of these places and human beings actions within these places. In America, people have more aesthetic appreciation towards wilderness while neglecting the appreciation of cultivated land. Their appreciation of wilderness is not deeply connected to the history of American Indians who had deep interaction with the land. I argue that through intercultural dialogue between Emerson and Chinese philosophy, we can create a new interpretation of place. With the expansion of places, human beings can enrich their understanding of themselves. When the self is connected to these places, conservation of these areas will become more effective.

References

1. Emerson, *Essays and Lectures*, ed. Joel Porte (The Library of America, 1983).
2. *Constructions of the Self*, ed. George Levine (New Brunswick, New Jersey: Rutgers University Press, 1992)
3. Robert D. Richardson, *Emerson: The Mind on Fire* (University of California Press, 1995).
4. Emerson, *The Portable Emerson*, ed. Jeffrey S. Cramer (Penguin Books, 2014).
5. Chan, Wing-tsit, *A Source Book in Chinese Philosophy*, trans. Wing-tsit Chan (Princeton University Press, 1963).
6. Yu, Ying-shi (余英时), 《论天人之际:中国古代思想起源试探》(*Between Heaven and Man: An Inquiry into the Origins of Ancient Chinese Thought*),Zhonghua Book Company, 2014.
7. Edward S. Casey, *Getting Back into Place* (Bloomington: Indiana University Press, 1993).
8. Holmes Rolston III, *Environmental Ethics: Duties to and Values in The Natural World* (Philadelphia: Temple University Press, 1988).

[11] Holmes Rolston III, *Environmental Ethics: Duties to and Values* in *The Natural World*, (Philadelphia: Temple University Press, 1988), pp. 112-113.

Part II
Nature

Although human beings live in a similar natural environment within the same Earth, their interpretation of nature, which is deeply influenced by philosophical and cultural traditions, differs greatly. This interpretation shapes divergent ethical and aesthetic attitudes toward nature. This chapter centers on the concept of nature through the lens of power and value. By value, I focus on the ultimate value in the metaphysical sense which is distinct from instrumental values that cast nature as a resource for human welfare. Both Emerson's conception of nature and the Chinese notion of heaven and earth position nature as a bearer of ultimate value, expressed through truth, goodness, and beauty. Yet their explanation of these ultimate values has great differences. This analysis unfolds across three sections: the concept of nature, the interplay between nature and value, and the contrasting perspectives of Emerson and Chinese philosophy on these themes. Emerson's approach reveals transcendence, while Chinese philosophy embodies immanence—an immanence that enfolds a form of transcendence, uniquely interpreted through the fusion of the two.

This chapter opens with a comparative examination of Emerson and Chinese philosophy, beginning with the concept of nature. To frame this comparison, it is necessary to consider the translation and evolution of the concept of "nature" in Chinese thought. Professor Wang Zhongjiang, in his work *Nature and Human: A Genealogical Study of Two Ideals in Modern China*,[1] offers a meticulous study of this concept's translational history, providing profound insights into its resonance within Chinese culture. Consulting early English-Chinese dictionaries, he catalogs various renderings of the Western notion of nature in modern China: "性" (xing, denoting inherent properties of men and things), "天地" (tiandi, encompassing the system of nature, the world, or the cosmic mechanism), "性理" (xingli, principles of nature), "天地运气" (tiandi yunqi, the course of nature), and "天理" (tianli, heavenly principles), among others. Wang notes that by the late Qing Dynasty and early Republic of China, influenced by Japanese lexicography, dictionaries began adopting "自然" (ziran) to define "nature." He cites the Japanese dictionary Shin Jiga (新尔

[1] Wang, Zhongjiang, *Nature and Human: A Genealogical Study of Two Ideals in Modern China*, The Commercial Press, 2018.

雅), which employed terms like "自然科学" (natural science), "自然力" (natural force), and "自然哲学" (natural philosophy), signaling a concept that integrates entities and scientific inquiry. This translational lens reveals that "天地" (heaven and earth) in Confucian and Daoist philosophy most closely aligns with Emerson's "nature," as it embraces all natural phenomena. Yet the understanding of these "all things" within "天地" fundamentally diverges from Emerson's vision. To illuminate this distinction, we turn to their metaphysical interpretations of nature and "天地," unpacking their respective conceptions of the material world.

Chapter 5
Nature in Emerson: Power and Values

The core categories Emerson uses to express the concept of nature are *natura naturans* and *nature naturta*. The former refers to the dynamic nature and the latter refers to the generated nature. In this chapter, I will use the concepts of power and values to analyze Emerson's concept of nature. These two concepts can help us gain a deeper understanding of Emerson's idea of self, virtue, and the pursuit of meaning in life through nature. In the following, I will examine Emerson's concept of nature through three aspects: (1) nature's creative and transformative power; (2) the ultimate values in nature and its relationship with power; (3) the scientific dimension of power.

1 Nature: Creative and Transformative Power

Emerson frequently uses the term "power" to explain nature. By power, he refers to creative and transformative power of nature. The two forms of power are interconnected. For him, nature has the power to create various forms which are the foundations of material nature. Forms are always in the process of transformation.

First, in terms of the creative power, Emerson's two foundational metaphysical principles, articulated in his 1857 journal entry, provide the logical starting point for understanding the meaning. In his journal, Emerson writes, "My philosophy holds to a few laws, 1. *Identity*, whence comes the fact that *metaphysical faculties & facts are the transcendency of physical*. 2. Flowing, or transition, or shooting the gulf, the perpetual striving to ascend to a higher platform, the same thing in new & higher forms."[1] These principles—identity and fluidity—underpin his conception of power. For Emerson, the force propelling the movement of matter is a metaphysical fact that transcends physical embodiment. This ultimate metaphysical fact possesses a

[1] Emerson, *The Journals and Miscellaneous Notebooks of Ralph Waldo Emerson*, vol. 14. ed. William H. Gilman, et al. (Cambridge, MA: Harvard University Press, 2011), pp. 191-192.

S. Gao, *Comparative Studies Between Emerson and Chinese Philosophy*, Nature Aesthetics: Interdisciplinary and Intercultural Dialogue,
https://doi.org/10.1007/978-981-95-0421-3_5

stable identity, which he expresses through diverse yet interrelated categories such as dynamic nature, thought, the Oversoul, and God—all frequently associated with the term "power". In describing dynamic nature, Emerson employs concepts like thought, spiritual principles, reason, and supreme nature. For instance, he declares, "Thought! It is the thread on which the system of nature and the heaven of heavens are strung. All that we see, or touch, or experience came from it, as leaves from a tree. The world, the galaxy, is but a scrap before the metaphysical power."[2] This passage positions thought as the originating metaphysical force, the ultimate cause of all existence. Thought (or Oversoul, God) has creativity. It creates different forms, similar to Plato's ideas. He writes, "But taking timely warning, and leaving many things unsaid on this topic, let us not longer omit our homage to the Efficent Nature, *natura naturans*, the quick cause, before which all forms flee as the driven snows, itself secret, its works driven before it in flocks and multitudes, (as the ancient represented nature by Proteus, a shepherd,) and in undescribable variety. It publishes itself in creatures, reaching from particles and spicula, through transformation on transformation to the highest symmetries, arriving at consummate results without a shock or a leap."[3] Thought in the above passage is often interchangeable with other concepts such as God, *natura naturans*, reason, supreme nature and so on. Thought does not directly produce material nature, rather it produces various forms which is similar to Plato's concept of ideas. From the above, we can see that Emerson's idea of creative power stresses the creation of forms.

Secondly, Emerson's concept of nature has the meaning of transformative power. This transformative power manifests as a continual evolution of forms, whereby the myriad forms generated by nature's dynamism perpetually advance toward a higher, more refined state. This transformative power is articulated through terms such as "tendency", "transformation" and "overflows". He frequently uses vivid literary language to depict this process of change in his essay titled History. For instance, he writes, "Upborne and surrounded as we are by this all-creating nature, soft and fluid as a cloud or the air..."[4]; "Nature is a mutable cloud which is always and never the same. She casts the same thought into troops of forms, as a poet makes twenty fables with one moral."[5]; "Nature is an endless combination and repetition of a very few laws. She hums the old well-known air through innumerable variations."[6] Through these metaphors, Emerson portrays nature as an entity in perpetual flux, which is driven by an inherent tendency toward transformation—a tendency underpinned by spiritual laws.

---

[2] Emerson, *The Later Lectures of Ralph Waldo Emerson*, vol. 2, ed. Ronald A. Bosco and Joel Myerson (Athens and London: University of Georgia Press, 2001), p. 77.

[3] Ralph Waldo Emerson, *Essays and Lectures*, ed. Joel Porte (New York: Literary Classics of the United States, 1983), p. 546.

[4] Ralph Waldo Emerson, *The Portable Emerson*, ed. Jeffrey S. Cramer (Penguin Books, 2014), p. 135.

[5] Ibid.

[6] Ibid., p. 136.

This spiritual foundation for the transformative power is evident in his assertion. He writes, "Every natural fact is a symbol of some spiritual fact."[7] For Emerson, all of the change of natural things and natural facts are based on the transformation of forms of nature. He envisions nature as ceaselessly generating new life forms. Emerson describes this incessant formal evolution as "overflow", a concept he explores in depth in his essay "The Method of Nature". He writes, "The method of nature: who could ever analyze it? That rushing stream will not stop to be observed. We can never surprise nature in a corner; never find the end of a thread; never tell where to set the first stone. The bird hastens to lay her egg: the egg hastens to be a bird. The wholeness we admire in the order of the world, is the result of infinite distribution. Its smoothness is the smoothness of the pitch of the cataract. Its permanence is a perpetual inchoation. Every natural fact is an emanation, and that from which it emanates is an emanation also, and from every emanation is a new emanation. If anything could stand still, it would be crushed and dissipated by the torrent it resisted, and if it were a mind, would be crazed; as insane persons are those who hold fast to one thought and do not flow with the course of nature."[8] This passage illuminates the essence of nature's creativity: an eternal process wherein new forms are incessantly produced, each inheriting and extending the legacy of its predecessors. Far from being isolated, these forms participate in a continuous chain of transformation.

Emerson's notion of transformative power aligns with his broader metaphysical framework, particularly his principle of fluidity. As he notes in his journals, "Step by step we scale this mysterious ladder; the steps are actions, the new prospect is power."[9] This upward trajectory reflects the teleological dimension of nature's dynamism, a process not merely of change but of progressive refinement. To further illustrate this, we might consider his observation in his essay "Circles", "Our life is an apprenticeship to the truth that around every circle another can be drawn; that there is no end in nature, but every end is a beginning; that there is always another dawn risen on mid-noon, and under every deep a lower deep opens."[10] Here, the interplay of repetition and renewal underscores the transformative power's capacity to sustain continuity amid perpetual innovation.

2 Nature: Power and Values

Emerson's conception of nature is intimately tied to values, which in turn are intricately linked to power. The logical relationship among nature, power, and values can be expressed as follows: (1) *natura naturans* as the productive aspect of nature is

[7] Ibid., p. 28.

[8] Ralph Waldo Emerson, *The Best Read Naturalist*, ed. Michael P. Branch&Clinton Mohs (University of Virginia Press, 2017), p. 134.

[9] Ralph Waldo Emerson, *The Portable Emerson*, ed. Jeffrey S. Cramer (Penguin Books, 2014), p. 228.

[10] Ibid., p. 226.

the bearer of ultimate value; (2) Value can generate creative power which causes the production of material nature, which is termed as *natura naturata* by Emerson. By values, I refer to the ultimate values that Emerson stresses in his work, which are truth, goodness, and beauty. The value dimension of nature can be best expressed by Emerson's core term "ecstasy". In his essay "The Method of Nature", he writes, "Nature can only be conceived as existing to a universal and not to a particular end; to a universe of ends, and not to one,—a work of *ecstasy*, to be represented by a circular moment, as intention might be signified by a straight line of definite length."[11] The term "ecstasy" denotes nature's purpose and combination of values. In the following, I will discuss each value and address the relationship between values and power.

First, truth is one of the three ultimate values for Emerson. In terms of truth, Emerson writes, "Once inhale the upper air, being admitted to behold the absolute natures of justice and truth, and we learn that man has access to the entire mind of the Creator, is himself the creator in the finite."[12] The word "truth" here has a strong value connotation. It represents an ultimate value. There is a flow direction in the concept of truth. It directs us from the visible world to invisible one, that is, the first cause. When Emerson explains truth, he often explains this concept with the concept of power. He says, "Thus, the first quality we know, in matter is, Centrality, which we commonly call gravity, and which holds the Universe together; which remains pure, and entire, and indestructible in each mote, as in masses and planets, and from each atom rays out illimitable influence. To this central essence answers Truth, in the intellectual world, Truth, whose centre is everywhere and its circumference nowhere, whose existence we cannot disimagine;"[13] From this quote, we can see that Emerson employ the scientific concept of power (gravity) to interpret the concept of truth. Truth as the value term can generate material powers. He writes, "Every truth tends to become a power. Every idea from the moment of its emergence, begins to gather material forces, and, after a little while makes itself known in the spheres of politics and commerce. It works first on thoughts, then on things, it makes feet, and afterward shoes; first, hands; then, gloves; makes the men, and so the age and its *materiel* soon after."[14] From this quote, it can be seen that in Emerson, truth as spiritual elements has a power of creation. The value of truth can generate material power without relying on any material substance.

Second, goodness is most important value among truth, goodness and beauty. Emerson doesn't give a definition of goodness. Instead, he describes the concept of goodness in terms of power. He states, "All the powers and rewards of Faith, which we find in the *Good*, hold equally in the region of the *True*. Integrity

---

[11] *The Complete Sermons of Ralph Waldo Emerson*, ed. Albert J.von Frank, et al. vol 4 (Columbia: University of Missouri Press, 1989-1992), pp.166-167.

[12] Ralph Waldo Emerson, *The Portable Emerson*, ed. Jeffrey S. Cramer (Penguin Books, 2014), pp. 47-48.

[13] Ralph Waldo, *The Best Read Naturalist*, ed. Michael P. Branch and Clinton Mohs (University of Virginia Press, 2017), p. 215.

[14] Ibid., p. 180.

is really the fountain of power, in one as in the other."[15] "Nothing divine dies. All good is eternally reproductive."[16] From the two passages, we can see that Emerson's idea of goodness has the following meanings: (1) Goodness has its own power for self-existence without depending on any other entities; (2) Goodness has the moral dimension. Emerson sometimes uses the moral term "justice" to express the idea of Goodness; (3) Goodness ensures the order of the universe. It is not something human beings choose to have rather it is an objective existence. Human beings can reveal its existence through good work but cannot establish the standard of goodness through choice.

Emerson's concept of goodness resonates with Plato's, particularly in their shared emphasis on goodness as the ultimate reality. For Plato, goodness transcends all beings, endowing knowledge with truthfulness and ensuring its validity. Emerson, however, extends Plato's framework into what he terms a "new metaphysics"—his Doctrine of Correspondence. Unlike Plato, who deems the natural world a mutable, imperfect reflection of the eternal Ideas, Emerson asserts the ontological reality of natural world. The world of nature as the effect of God is real rather than illusion. He argues that while nature is in constant flux, the spiritual law governing this change remains immutable. This law constitutes the truth of the natural world, corresponding directly to the human mind. Consequently, understanding nature becomes a pathway to self-knowledge. Emerson credits science with revealing this truth but insists that scientists must overcome linguistic barriers to make it accessible to the public. He places greater emphasis on poets, whose language—rooted in scientific insight yet transcending its arid precision—serves as a vital conduit for conveying nature's truth.

For a better understanding of the relationship between values and power in Emerson, I would like to mention Baruch Spinoza's concept of nature and its relationship with values and power. Spinoza, similar to Emerson, also uses two terms to express the concept of nature: *natura naturans* and *natura naturata.* The former refers to the causing creative power and the latter refers to the physical nature, which is the result of causing power. For the causing power, he also uses the concept of God to express this idea. He says, "From this it follows that God is not only the cause of things beginning to exist, but also of their persevering in existing, *or* (to use a Scholastic term) God is the cause of being of things."[17] This is an important point I would like to stress here about Spinoza's concept of *natura naturans* and its relationship with value and power. Spinoza often uses mathematics and geometry to express the creative power of *natura naturans.* Mathematics and geometry as scientific truth doesn't have ultimate ends and it doesn't have value dimension.

From the above comparison, it can be seen that the core difference between Emerson and Spinoza in their understanding of nature's creative power lies in the

---

[15] Ibid., p. 179.

[16] Ralph Waldo Emerson, *The Portable Emerson*, ed. Jeffrey S. Cramer (Penguin Books, 2014), p. 26.

[17] Benedict de Spinoza, *A Spinoza Reader: The Ethics and Other Works*, trans. Edwin Curley (Princeton University Press, 1994), p. 102.

idea of value. Emerson's concept of nature and power is closely related to the ultimate values of truth, goodness and beauty. He contends that natureembodies values of truth, goodness and. These values inevitably manifest through "revelation." He writes, "We distinguish the announcements of the soul, its manifestations of its own nature, by the term *Revelation*. These are always attended by the emotion of the sublime. For this communication is an influx of the Divine mind into our mind."[18] For Emerson, God's existence is thus revealed through tangible creations, apprehended not through rational proof but via unconditional faith. This aligns Emerson's thought with Plotinus, who identifies the ultimate entity as "the One"—the primal cause of all generation and change. Plotinus employs terms like "the first" and "the source of all things" to describe the One, emphasizing its creativity as an inherent trait of its perfection.[19] He asserts, "It is precisely because there is nothing within the One that all things are from it: in order that Being may be brought about, the source must be no Being but Being's generator, in what is to be thought of as the primal act of generation. Seeking nothing, possessing nothing, lacking nothing, the One is perfect and, in our metaphor, has overflowed, and its exuberance has produced the new: this product has turned again to its begetter and been filled and has become its contemplator and so an Intellectual-Principle."[20] For Plotinus, perfection necessitates creation, yet the One remains static, permeating its image without motion.

Emerson mirrors Plotinus in linking creativity to perfection, which he too roots in value rather than reason. The Plotinus states in his *Enneads*, "We are in search of unity; we are to come to know the principle of all, the Good and First; therefore we may not stand away from the realm of Firsts and lie prostrate among the lasts: we must strike for those Firsts, rising from things of sense which are the lasts."[21] For Plotinus, good is a value term which has moral dimension and it transcends all material things. Emerson echoes this in the essay titled Spiritual Laws, "Place yourself in the middle of the stream of power and wisdom which animates all whom it floats, and you are without effort impelled to truth, to right and a perfect contentment."[22]

Third, the final value is beauty. Unlike Kant, who defines beauty in his *Critique of the Power of Judgment*, Emerson conceptualizes beauty in a different way. A deeper understanding of his concept of beauty can be gained by examining the two key terms he repeatedly associates with beauty: transition and necessity. He writes, "Beauty is the moment of transition, as if the form were just ready to flow into other forms.";[23] "Beauty rests on necessities."[24] Emerson's concept of beauty has two meanings: (1)

---

[18] Ralph Waldo Emerson, *The Portable Emerson*, ed. Jeffrey S. Cramer (Penguin Books, 2014), p. 217.

[19] Plotinus, *Enneads*, trans. Stephen MacKenna (Penguin Classics, 1991), p. 361.

[20] Ibid.

[21] Ibid., p. 538.

[22] Ralph Waldo Emerson, *The Portable Emerson*, ed. Jeffrey S. Cramer (Penguin Books, 2014), p. 195.

[23] Ralph Waldo Emerson, *Essays and Lectures*, ed. Joel Porte (New York: Literary Classics of the United States, 1983), p. 1105.

[24] Ibid., p. 1106.

It refers to the ultimate value and also the feature of spiritual law which is immanent in nature. The beauty is based on the truth in nature. Transition is one aspect of truth; (2) It is used to describe the beauty of physical nature. The first meaning of beauty is paramount since it determines the expressions of the beauty of physical nature. In this aspect, Joseph Urbas makes an insightful remark. "But the forms of 'nature passive' are not *themselves* the ultimate cause of our feeling of wonder, even when we savor the beauty of the material landscape in all its variety and particularity. The real cause is the spirit creative that is present and alive, ..."[25] Nature passive is effect of efficient nature and for Emerson, it doesn't have intrinsic value for their own sake and their beauty is not related to any properties rather is related to the creative spirit, which is the beautiful principle on the necessity.

Among the ultimate values of truth, goodness, and beauty, which value holds the highest rank, and what is their logical interrelation? This inquiry is crucial, as it informs Emerson's interpretation of aesthetic appreciation of nature. Emerson stresses that goodness occupies the preeminent position, with truth and beauty as its subordinates. Emerson's metaphysical concept of goodness inherits Plato's concept of Goodness. However, he further develops Plato's concept of Goodness. Plato positions goodness above truth and knowledge, as an entity transcending existence. In *The Republic*, he writes,"Therefore, say that what provides the truth to the things known and gives the power to the one who knows, is the *idea* of the good. And, as the cause of the knowledge and truth, you can understand it to be a thing known; but, as fair as these two are—knowledge and truth—if you believe that it is something different from them and still fairer than they, your belief will be right. As for knowledge and truth, just as in the other region it is right to hold light and sight sunlike, but to believe them to be sun is not right; so, too, here, to hold these two to be like the good is right, but to believe that either of them is the good is not right. The condition which characterizes the good must receive still greater honor."[26] From this quote, we can see that Plato's concept of goodness has two features. (1) Goodness transcends daily life, surpassing knowledge, truth, and beauty; (2) It endows truth and knowledge with veracity, illuminating the intelligible realm. Thus, goodness is the ultimate source of the world of Ideas, exceeding beauty.

Emerson aligns with Plato in prioritizing goodness, as seen in his essay "Circle": "Yet this incessant movement and progression which all things partake could never become sensible to us but by contrast to some principle of fixture or stability in the soul. Whilst the eternal generation of circles proceeds, the eternal generator abides. That central life is somewhat superior to creation, superior to knowledge and thought, and contains all its circles. Forever it labors to create a life and thought as large and excellent as itself, but in vain, for that which is made instructs how to make a better"[27] Here, the circle's center—God—represents goodness as a fixed principle, eternally

---

[25] Joseph Urbas, *The Philosophy of Ralph Waldo Emerson* (Routledge, 2021), p. 160.

[26] Plato, *The Republic of Plato*, trans. Allan Bloom (New York: Basic Books, 1968), p. 189.

[27] Ralph Waldo Emerson, *The Portable Emerson*, ed. Jeffrey S. Cramer (Penguin Books, 2014), p. 235.

generating life while encompassing all within its power, though created forms strive to transcend themselves without escaping its orbit.

In terms of the concept of goodness, neither Plato nor Emerson defines it explicitly, but both associate it with order. In *Laws*, Plato asserts, "First, there is the evidence of the earth, the sun, the stars, and all the universe, and the beautiful ordering of the seasons, marked out by years and months; and then there is the further fact that all Greeks and barbarians believe in the existence of gods."[28] In *Phaedo*, Socrates critiques Anaxagoras' materialist wisdom, seeking instead "what is best" for each thing. He elaborates in *Philebus*: "Therefore if we are unable to net the good in a single concept, we must use three to capture it, namely beauty, proportion and truth. This is our position: supposing these three to be a single unit, we could not go wrong if we held it responsible for the components of the mixture, in the sense that, *qua* good, it makes the mixture good too."[29] For Plato, goodness integrates beauty, order, and truth, rendering the mixture good. In his *Republic*, he illustrates this with astronomy, "don't you suppose that a man who is really an astronomer will have the same persuasion in looking at the movements of the stars? He will hold that the craftsman of heaven composed it and what's in it as beautifully as such works can be composed. But as for the proportion of night to day, of these to a month, of a month to a year, and of the rest of the stars to these and to one another, don't you think he will consider strange the man who holds that these are always the same and deviate in no way at all? For these things are connected with body and are visible. Hence won't he consider it strange to seek in every way to grasp their truth?"[30] Plato's goodness manifests as a mathematical order, deemed beautiful and good.

This Platonic aesthetic, prioritizing geometric proportion, shaped natural aesthetics negatively, as Eugene C. Hargrove observes, "Aesthetically, Plato's notion of perfection translates into a preference for mathematical proportions and ideal geometric shapes. This kind of perfection is responsible for much of the sixteenth- and seventeenth-century distaste for mountainous scenery, which for a time inhibited the development of modern nature appreciation."[31] He mentions Thomas Burnet, who was deeply influenced by Plato, despaired at the Alps' irregularity: "Thomas Burnet, for example, infected by such a Platonic aesthetic perspective, came close to losing his faith in God when, arriving in the vicinity of the Alps in 1671, he found that he was unable to discern any regularity, proportion, or symmetry in the slopes and shapes of the mountains before him."[32] Emerson, however, departs from this emphasis on mathematical order. He asserts, "Whatever theology or philosophy we rest in, or labor after, the students of Nature have all agreed that in Nature nothing

---

28 Plato, *Laws,* vol. XI of *Plato in Twelve Volumes*, trans. R. G. Bury (Harvard University Press), p. 301.

29 Plato, *Philebus*, trans. Robin A. H. Waterfield (Penguin Books, 1982), p. 145.

30 Plato, *The Republic of Plato*, Second Edition, trans. Allen Bloom (Basic Books, 1968), p. 209.

31 Eugene C. Hargrove, *Foundations of Environmental Ethics* (*Environmental Ethics Books*, 1989), p. 28.

32 Ibid.

is false or unsuccessful." [33] For Emerson, all of God's creations, guided by ultimate value, are inherently beautiful, irrespective of geometric perfection.

3 Nature: Scientific Dimension

Emerson emphasizes science when explaining the concept of nature with its relationship with power and value. Eric G. Wilson, in his book titled *Emerson's sublime Science*, meticulously examines the scientific undertones of Emerson's deployment of "power" and traces Coleridge's influence on this conceptual framework. Wilson highlights that in 1829, during his engagement to Ellen Tucker and shortly before he resigned from the church in 1830, Emerson was deeply immersed in James Marsh's American edition of Coleridge's *Aids to Reflection.* In this work, Coleridge asserts, "That what you see is blood, is flesh, is itself the work, or shall I say, the translucence, of the invisible energy, …'"[34] This proposition reflects Coleridge's confidence in the dynamic nature of physical phenomena, a perspective bolstered by the experimental findings of his scientist contemporaries, electrochemist Humphry Davy and electromagnetist Hans Christian Orsted. Their work demonstrated that organic structures are perpetually in flux, driven by an underlying spiritual force. He defines science as follows, "Science is the arrangement of the phenomena of the world after their essential relations. It is the reconstruction of nature in the mind. This is at once its ideal and its historical aspect. The most striking trait of modern science is its approximation towards central truths. On all sides, it is simplifying its laws and finding one cause for many effects."[35] He emphasizes the cognition of truth brought by science to individuals. The science understood by Emerson is the reconstruction of nature in the human mind. He believes in science's approach to core truths and that science can simplify various laws and find the ultimate cause of phenomena. Emerson's descriptions of nature in many prose works are scientific. For example, he writes, "Genius detects through the fly, through the caterpillar, through the grub, through the egg, the constant individual, through countless individuals the fixed species; through many species the genus; through all genera the steadfast type; through all the kingdoms of organized life the ternal unity."[36] The concepts of "species" and "class" in this passage indicates that Emerson uses knowledge of ecology to explain nature. However, one thing I would like to stress here is that Emerson also stresses joyful

[33] Ralph Waldo Emerson, *The Best Read Naturalist*, ed. Michael P. Branch and Clinton Mohs (University of Virginia Press, 2017), p. 61.

[34] Samuel Taylor Coleridge, *Aids to Reflection*, ed. Henry Nelson Coleridge (New York: Swords, Stanford & co, 1839), p. 313.

[35] Ralph Waldo Emerson, *The Best Read Naturalist*, ed. Michael P. Branch and Clinton Mohs (University of Virginia Press, 2017), p. 117.

[36] Ralph Waldo Emerson, *The Portable Emerson*, ed. Jeffrey S. Cramer (Penguin Books, 2014), p. 135.

science. "Science was false by being unpoetical."[37] For him, there are some negative aspect of science. Scientists sometimes only focus too much on details of individual animals or plants and ignore their relationship with the Soul.

References

1. Emerson, *The Best Read Naturalist*, ed. Michael P. Branch&Clinton Mohs (University of Virginia Press, 2017.
2. Emerson, *The Journals and Miscellaneous Notebooks of Ralph Waldo Emerson*, vol. 14. ed. William H. Gilman, et al. (Cambridge, MA: Harvard University Press, 2011).
3. Emerson, *The Later Lectures of Ralph Waldo Emerson*, vol. 2, ed. Ronald A. Bosco and Joel Myerson (Ahens: University of Georgia Press, 2001).
4. Ralph Waldo Emerson, *The Portable Emerson*, ed. Jeffrey S. Cramer (Penguin Books, 2014).
5. Ralph Waldo Emerson, *Letters and Social Aims in Emerson's Complete Works*, vol. VIIII (The Riverside Press, 1884).
6. *The Complete Sermons of Ralph Waldo Emerson*, ed. Albert J.von Frank, et al. vol 4 (Columbia: University of Missouri Press, 1989–1992).
7. Spinoza, *A Spinoza Reader: The Ethics and Other Works*, trans. Edwin Curley (Princeton University Press, 1994).
8. Plotinus, *Enneads*, trans. Stephen MacKenna (Penguin Classics, 1991).
9. Emerson, *The Collected Works of Ralph Waldo Emerson*, vol. 1, ed. Alfred R. Ferguson (Harvard University Press, 1971).
10. Plato, *Plato in Twelve Volumes, Volume II*, XI Laws, trans. R. G. Bury (Cambridge, Massachusetts: Harvard University Press, 1926).
11. Plato, *Philebus*, trans. Robin A. H. Waterfield (Penguin Books, 1982).
12. Plato, *The Republic of Plato*, trans. Allan Bloom (Basic Books, 1968).
13. Eugene C. Hargrove, *Foundations of Environmental Ethics* (*Environmental Ethics Books*, 1989).

[37] Ralph Waldo Emerson, *Letters and Social Aims in Emerson's Complete Works*, vol. VIII (The Riverside Press, 1884), p. 15.

Chapter 6
Heaven and Earth in Confucianism and Daoism: Power and Values

In Confucianism and Daoism, there is no corresponding Emerson's concept of nature. The concept closest to Emerson's nature is Heaven and Earth. Like Emerson's concept of nature, the concept of Heaven and Earth is also deeply connected to power and value. However, the meaning of the concepts of power and value differ greatly from Emerson's concept of power and value. In Confucianism and Daoism, power does not have the scientific dimension as Emerson interprets. The concept of power is related to the concept of vital energy, which is termed as *qi*. Qi is deeply related to value. In the following, I will elaborate this connection of concepts among Heaven and Earth, power and value.

1 The Concept of Heaven and Earth in Confucianism

Before addressing the relationship among Heaven and Earth, power, and value, it is necessary for us to take a look at the passages which are related to Heaven and Earth from *The Analects*:

> The Master said, 'At fifteen I set my heart on learning; at thirty I took my stand; at forty I came to be free from doubts; at fifty I understood the Decree of Heaven; at sixty my ear was atuned; at seventy I followed my heart's desire without overstepping the line.'[1]

> Wang-sun Chia said, 'Better to be obsequious to the kitchen stove Than to the south-west corner of the house. What does that mean?' The Maser said, 'The saying has got it wrong. When you have offended against Heaven, there is nowhere you can turn to in your prayers.'[2]

> The border offical of Yi requested an audience, saying, 'I have never been denied an audience by any gentleman who has come to this place.' The followers presented him. When he came out, he said, 'What worry have you, gentlemen, about the loss of office? The Empire has

[1] Confucius, *The Analects*, trans. D. C. Lau (Penguin Books, 1979), p. 63.

[2] Ibid., p. 69.

S. Gao, *Comparative Studies Between Emerson and Chinese Philosophy*, Nature Aesthetics: Interdisciplinary and Intercultural Dialogue,
https://doi.org/10.1007/978-981-95-0421-3_6

long been without the Way. Heaven is about to use your Master as the wooden tongue for a bell.'[3]

Tzu-kung said, 'One can get to hear about the Master's accomplishments, but one cannot get to hear his views on human nature and the Way of Heaven.'[4]

The Master said, 'Heaven is author of the virtue that is in me. What can Huan T'ui do to me?'[5]

Confucius said, 'There are three things the gentleman should guard against. In youth when the blood and ch'i are still unsettled he should guard against the attraction of feminine beauty. In the promise of life when the blood and ch'i have become unyielding, he should guard against bellicosity. In old age when the blood and ch'i have declined, he should guard against acquisitiveness.'[6]

The Master went to Wu Ch'eng. There he heard the sound of stringed instruments and singing. The Master broke into a smile and said, 'Surely you don't need to use an ox-knife to kill a chicken.' Tzu-yu answered, 'Some time ago I heard it from you, Master, that the gentleman instructed in the Way loves his fellow men and that the small man instructed in the Way is easy to command.' The Master said, 'My friends, what Yen says is right. My remark a moment ago was only made in jest.'[7]

Feng Yu-lan provides a comprehensive summary of the Confucian concept of heaven in his seminal work titled *A History of Chinese Philosophy* [8]. He points out that the conception of Heaven or *T'ien* has five meanings. Heaven in Chinese philosophy is material, ruling, fatalistic, naturalistic and ethical. His taxonomy captures the breadth of Heaven's connotations in Confucianism. However, he neglects the important dimension of Heaven, that is, its interrelationship with power and values. The five meanings of Heaven are all connected to the concept of power. In terms of power, it is not related to scientific worldview as in Emerson's lifetime, rather this power is related to Chinese philosophical concepts of *yin* and *yang*.

The concept of Heaven and Earth in Confucianism has the meaning of creative power. It is always in a process of creating life. One point I would like to stress here is that creative power of Heaven and Earth is not derived from reason or divine consciousness. Rather, it is the result of the interaction of *yin* and *yang*. The creative power of Heaven and Earth is intimately related to values. In Confucianism, the Dao is immanent in this generative process. Dao is not a transcendental entity but the spontaneous way of reality itself. Dao is a value-laden term. Its value lies in moral goodness and for Confucianism, moral goodness refers to benevolence. The life-creating power of Heaven and Earth understood by Confucianism has mystery and inevitability. The concept of Heaven in Confucianism is often used in conjunction with the concept of "fate". Fate refers to an irresistible force that can dominate human life, happiness, longevity, human emotions, and changes in nature. This force is

---

[3] Ibid., p. 71.

[4] Ibid., p. 78.

[5] Ibid., p. 89.

[6] Ibid., p. 140.

[7] Ibid., p. 319.

[8] Youlan Feng, *A History of Chinese Philosophy*, trans. Derk Bodde (London, 1952), p. 31.

inevitable and mysterious. Fate encompasses a situational nature, which is a situation brought about by various forces of the time. Under this situation, the occurrence of some events is inevitable and cannot be resisted.

The essential difference between Emerson and Confucianism lies in their understanding of life. This difference can be examined by comparing the concept of *yin* and *yang* in Chinese philosophy mentioned by Emerson. Emerson also mentioned the idea of *yin* and *yang* in his essay, but the idea of *yin* and *yang* he mentioned is essentially different from the concept of *yin* and *yang* in Chinese philosophy. Let us first take a look at the concept of *yin* and *yang* mentioned by Emerson in his essay "Compensation":

> Polarity, or action and reaction, we meet in every part of nature; in darkness and light; in heat and cold; in the ebb and flow of waters; in male and female; in the inspiration and expiration of plants and animals; in the equation of quantity and quality in the fluids of the animal body; ...An inevitable dualism bisects nature, so that each thing is a half, and suggests another thing to make it whole; as, spirit, matter; man, woman; odd, even; subjective, objective; in, out, upper, under; motion, rest; yea, nay. ...The same dualism underlies the nature and condition of man. Every excess causes a defect; every defect an excess. ...The true doctrine of omnipresence is that God reappears with all his parts in every moss and cobweb. The value of the universe contrives to throw itself into every point. If the good is there, so is the evil; if the affinity, so the repulsion; if the force, so the limitation. Thus is the universe alive. All things are moral. That soul, which within us is a sentiment, outside of us is a law. We feel its inspiration; out there in history we can see its fatal strength.[9]

Emerson's idea of polarity also contains the characteristics of mutual opposition, complementarity, and balance. But in Emerson, what is more important than the existence of opposing sides is the ultimate spiritual law that brings these oppositions. It is the spiritual law that produces these oppositions. The spiritual law itself is alive, and it is superior to material life. This is a major difference between Emerson and Chinese philosophy. In Chinese philosophy, *yin* and *yang* are not produced by spiritual law and the interactions are spontaneous without following any spiritual law. *Yin* and *yang* is made up of *qi* which is both material and spiritual. There is no hierarchy between them.

2 Heaven and Earth in Confucianism: Power and Value

In Confucianism, Heaven and Earth is full of creative power, which is deeply connected to value. By creative power, I refer to Heaven and Earth's unending process of creating life. This creation is not based on perfection revealed from mathematics as Spinoza explained, but is based on the vital energy (*qi*) which is closely related to value. Confucianism uses different moral categories to explain this value. In terms of these moral categories, different philosophers emphasize different aspects, with the

---

9 Ralph Waldo Emerson, *The Portable Emerson*, ed. Jeffery S. Cramer (Penguin Books, 2014), pp. 175-178.

Doctrine of the Mean emphasizing sincerity, Zhangzai emphasizing harmony, and Zhuxi emphasizing benevolence.

The moral term of sincerity appears many times in *the Doctrine of the Mean.*

> Sincerity means the completion of the self, and the Way is self-directing. Sincerity is the beginning and end of things.[10]
>
> Therefore absolute sincerity is ceaseless. Being ceaseless, it is lasting. Being lasting, it is evident. Being evident, it is infinite.[11]
>
> The Way of Heaven and Earth may be completely described in one sentence: They are without any doubleness and so they produce things in an unfathomable way.[12]

The concept of sincerity is often used in dealing with human moral relationships. It has the following meanings: (1) unselfish; (2) persistent effort; (3) affection of love. In *The Doctrine of the Mean*, sincerity (cheng) has the metaphysical dimension. By metaphysical dimension, I mean that sincerity is considered as ultimate reality and values. It has the ontological status, which is derived from its religious origin.

In this regard, Jinhua Jia's etymological research of the concept of sincerity in article *Writings, Emotions, and Obligations: The Religious-Ritual Origin of the Classical Confucian Conception of Cheng (Sincerity)*[13] offers valuable insights into the religious origin of metaphysical meaning of sincerity. According to her study, the meaning of sincerity can be traced to its religious origins and practices and the metaphysical meaning of sincerity is developed from them rooted in China's shamanistic traditions. This perspective aligns with views held by numerous scholars, Zehou Li in his book titled *The Origins of Chinese Thought: From Shamanism to Ritual Regulations and Humaneness*[14] argues that in shamanistic activities, sincerity constitutes the paramount emotional and spiritual element. Shamans, who historically held significant roles in feudal courts, were seen as the sole intermediaries capable of communing with the divine, thereby securing the legitimacy of political rule. Li posits that when a shaman's emotions reach a frenzied state—marked by visible physical manifestations—the gods are summoned forth, suggesting that Chinese deities "emerge through the shaman's dance."[15] Similarly, Jia Jinhua, through meticulous historical evidence and etymological analysis, demonstrates that sincerity was intricately tied to early Chinese religious rituals, serving as the essential emotional state for facilitating communication between humans and the divine, thus invoking blessings upon the human realm.

---

[10] *A Source Book in Chinese Philosophy*, trans. Wing-Tsit Chan (Princeton University Press, 1963), p. 108.

[11] Ibid., p. 109.

[12] Ibid.

[13] Jinhua Jia, "Writings, Emotions, and Obligations: The Religious-Ritual Origin of the Classical Confucian Conception of Cheng (Sincerity)." *Religions* 2021, 12 (6): https://doi.org/10.3390/rel12060382.

[14] Zehou Li, *The Origins of Chinese Thought: From Shamanism to Ritual Regulations and Humaneness*, trans. Robert A. Carleo III (Brill, 2008).

[15] *A Source Book in Chinese Philosophy*, trans.Wing-Tsit Chan (Princeton University Press, 1963), p. 108.

Over time, sincerity evolved from a subjective human experience to an ontological principle, representing the creative potency of Heaven and Earth. Despite this elevation, its logical connection to its original meaning persists, manifesting in two key aspects: (1) Sincerity exhibits an outward directionality. The life-generating process driven by sincerity is not merely a self-contained act of perfection but is oriented toward others. (2) Sincerity carries an emotional resonance, directed toward humanity and all things, imbued with a profound love for life. This life is not an abstract notion but a concrete, organic existence unfolding within the dimensions of time and space.

In human society, sincerity carries two principal connotations. It constitutes the core virtue of morally exemplary individuals encompassing sincerity toward oneself and sincerity toward others. In this sense, the virtue of sincerity entails not only personal self-cultivation but also the active promotion of moral improvement in others. This is similar to Kant's concept of moral perfection, but there is a fundamental difference. Here a comparative analysis between Kant's idea of promoting others' happiness and the Confucian idea of promoting others' virtue can help us gain a deep understanding of the moral concept of sincerity. According to Kant, we can not complete the morality of others, as moral perfection depends on the individual's own autonomous rationality and good will. However, we have an imperfect duty of beneficence to promote the happiness of others as an end in itself. In contrast, Confucianism emphasizes not only self-cultivation but also the active responsibility to help others develop their virtues. In addition, it is sincerity that gives meaning to all things. Without sincerity, even if there are concrete things, they lose their meaning.

Sincerity as moral value is closely related to value of truth and beauty. The truth referred to by Confucianism focuses on the real existence, the original appearance of all things. This real existence refers to the existence in the ontological sense. Confucianism does not strive to prove such existence like many Western thinkers, such as Descartes' efforts to prove the existence of God, but take its true existence for granted. The basis of this true existence is sincerity, which allows the true existence to be revealed. Without sincerity, existence is in a state of being obscured and will not become real. Sincerity is also the basis of beauty. In terms of the relationship between beauty and sincerity, it can be seen in the *Doctrine of the Mean*, "The Way of Heaven and Earth may be completely described in one sentence: They are without any doubleness and so they produce things in an unfathomable way. …*The Book of Odes* says, 'The Mandate of Heaven, how beautiful and unceasing.'"[16]

Song Confucianist Zhangzai employs the term of harmony to describe the Way of Heaven and Earth. He writes, "The Great Harmony is called the Way (Tao, Moral Law). It embraces the nature which underlies all counter processes of floating and sinking, rising and falling, and motion and rest. It is the origin of the process of fusion and intermingling, of overcoming and being overcome, and of expansion and contraction."[17] For Zhangzai, harmony is the result of opposing power of *yin* and

[16] *A Source Book in Chinese Philosophy*, ed. Wing-Tsit Chan (Princeton University Press, 1963), p.109.

[17] Ibid., p. 500.

yang. Harmony as the ontological principle will guide the interaction of *yin* and *yang*. Harmony is a value term in moral sense.

Zhuxi uses the term of benevolence (Jen) to express the moral goodness of Heaven and Earth. He writes, " 'The mind of Heaven and Earth is to produce things.' In the production of man and things, they receive the mind of Heaven and Earth as their mind. Therefore, with reference to the character of the mind, although it embraces and penetrates all and leaves nothing to be desired. nevertheless, one word will cover all of it, namely, *jen* (humanity)."[18] For Zhuxi, the natural process of generating life includes origination, flourishing, advantages, and firmness. It represents the four seasons of the year and also represents four cardinal virtues (benevolence, righteousness, propriety, and wisdom). Among the four virtues, benevolence (jen) is most important and it is the foundation of all the other three virtues. Jen has the ontological status and it controls the entire generating process of Heaven and Earth.

3 Heaven and Earth in Daoism: Power and Value

In Daoism, the concept of Heaven and Earth as in Confucianism is also deeply related to power and value. When Daoism explains power, it is similar to Confucianism that power is related to the concept of *qi*. However, when Daoism explains the relationship between power and value, it differs greatly with Confucianism. For Daoism, it stresses the value of spontaneity, which is the combination of truth, goodness, and beauty. In the following, I will mainly focus on the power of Heaven and Earth and its relationship with value in Zhuangzi since it is the core to explain Daoism' understanding of power. The basic relationship between power and value in Daoism is that value will influence the movement of power but value will not directly generate power. I will examine this relationship in three aspects: (1) the value of spontaneity; (2) value of spontaneity and its relationship with truth, good and beauty; (3) value of spontaneity and harmony.

First, spontaneity refers to being self-so. Self-so in Chinese is 自然(ziran), which is in essence the value term rather than the natural world. The concept of "so" in Chinese is ran 然, it is not value neutral term. In this aspect, I embrace Bo Wang's has provided a detailed etymology research and argues that it is indeed a value term.[19] Spontaneity is in opposite to man-made values. Zhuangzi contrasts the spontaneous order of Heaven and Earth with the purposeful pursuits of humanity. He illustrates this in the essay of Autumn Floods. "The River God said, 'What do you mean by the Heavenly and what by the Human? Ruo of the Northern Sea said, "That cows and horses have four legs is the Heavenly. The bridle around the horse's head and the ring through the cow's nose are the Human. Hence, it is said, 'Do not use the Human to destroy the Heavenly, do not use the purposive to destroy the fate, do not

[18] Ibid., pp. 593-594.

[19] Bo Wang, Re-discussion on "Ran" and "Nature" ("然" 与 "自然": 道家自然观念的再研究), *Philosophical Research*, 2018 (10).

sacrifice what you have attained [from Heaven]for the sake of mere names.' Hold onto this carefully, for then you can return to what is genuine in you."[20] Here, Heaven represents unadulterated nature, while human actions—such as taming animals—impose artificiality. Artificiality also refers to man-made values which are values towards the pursuit of knowledge, wealth, fame or power.

Zhuangzi writes, "To understand what is done by Heaven, and also what is done by Man, that is the utmost. To understand what is done by the Heavenly: just in being the Heavenly, as the way all beings are born, what it does is give birth to them all. To understand what is to be done by the Human: that would be to use what your understanding understands to nurture what your understanding does not understand. You could then live out all your natural years without being cut down halfway. And that would indeed be the richest sort of knowledge."[21] Human pursuits—knowledge, fame, morality—disrupt the natural flow, whereas heaven's truth lies in its lack of purpose. This truth, Zhuangzi's ultimate value, transcends conventional goodness, manifesting as ziran (nature). He writes in the following passage, "Do not try to develop what is natural to man; develop what is natural to Heaven. He who develops Heaven benefits life; he who develops man injures life. Do not reject what is of Heaven, do not neglect what is of man, and the people will be close to the attainment of Truth."[22] Unlike Emerson's nature, tied to causal laws, Zhuangzi's concept of ziran (nature) is not an entity but a value of spontaneous authenticity, beyond good/evil or beauty/ugliness, yet imbued with its own transcendent goodness.

Second, Heaven and Earth and core values. Zhuangzi ties Heaven and Earth to the broader framework of Dao, which is combination of truth, goodness, and beauty. He asserts, "'Heaven and Earth' just means the vastest among forms, 'yin and yang' just means the vastest among forces, and the 'Course' just means what is most unbiased among all activities."[23] In Tiandi, he elaborates, "Although Heaven and Earth are great, their transformation is equal; although there are many things, their governance is one; although there are many people, their lord is the king. The king originates from virtue and is established by heaven. Therefore, it is said: the ancient kings who ruled the world did nothing, they were just virtuous."[24] Dao precedes all, as he states in the essay titled *The Great Source as Teacher*: "Being its own root and its own foundation, it exists firmly even when heaven and earth are not yet there."[25] For Zhuangzi, Heaven and Earth has order and this order is beautiful. "Heaven and Earth possess vast beauties, but they do not speak of them. The four seasons have

---

[20] Zhuangzi: *The Essential Writings with Selections from Traditional Commentaries*, trans. Brook Ziporyn (Hackett Publishing Company, Inc., 2009), p. 73.

[21] Ibid., p. 39.

[22] *The Complete Works of Chuang Tzu*, trans. Burton Watson (New York: Columbia University Press), p. 199.

[23] Zhuangzi, *Zhuangzi: The Essential Writings with Selections From Traditional Commentaries* (Hackett Publishing Company, 2009), p. 73.

[24] Ibid., p. 82.

[25] Ibid., p. 43.

their unconcealed regularities, but they do not discuss them. Each of the ten thousand things makes its own perfect sense but does not explain it."[26]

Virtue (德)emerges as Dao's expression in living things, characterized by "emptiness, stillness, limpidity, silence, inaction," which Zhuangzi deems "the level of Heaven and earth".[27] He contrasts this with Confucian benevolence, noting, "The four seasons arrive one after the other"[28] and "For if I compare myself to all the creatures taking shape between heaven and earth and receiving vital energy from the yin and yang, I see that my position between Heaven and Earth is like that of a small stone or a tiny weed on a vast mountain."[29] This spontaneity, devoid of moral mission, reflects a virtue transcending human norms, aligning Heaven and Earth with Dao's natural rhythm.

Third, Heaven as a value standard for Human morality. Zhuangzi's Heaven also serves as a moral exemplar, guiding humanity toward authenticity. He writes, "The sage penetrates bafflement and complication, rounding all into a single body, yet he does not know why—it is his inborn nature. He returns to fate and acts accordingly, using Heaven as his teacher, and men follow after, pinning labels on him."[30] For Zhuangzi, the virtue of Heaven and Earth is naturalness, which serves as the guidance for the virtue of human beings. "This Teacher of mine, this Teacher of mine—he passes judgment on the ten thousand things but he doesn't think himself righteous; his bounty extends to ten thousand generations but he doesn't think himself benevolent. He is older than the highest antiquity but he doesn't think himself long-lived; he covers heaven, bears up the earth, carves and fashions countless forms, but he doesn't think himself skilled. It is with him alone I wander."[31] Heaven's virtue lies in its effortless unfolding, not imposed values.

Heaven's virtue provides guidance for human virtue. In the chapter titled Heaven and Earth, Zhuangzi writes, "Pervading Heaven and earth: that is the Way. Moving among the ten thousand things: that is Virtue. Superiors governing the men below them: that is called administration."[32] For Zhuangzi, Dao is closely related to virtue. Virtue nurtures everything and makes them flourish. Human beings' virtue is founded on Heaven's virtue. In terms of the relationship between Dao and virtue, it is best explained in his chapter titled Heaven and Earth, "In the Great beginning, there was nonbeing; there was no being, no name. Out of it arose One; there was One, but it has no form. Things got hold of it and came to life, and it was called Virtue."[33] From this passage, it can be seen that virtue gives all things life and helps them flourish. Without virtue, there will be no life. Human beings' virtue is based on this metaphysical sense

---

[26] Ibid., p. 86.

[27] Zhuangzi, *The Complete Works of Chuang Tzu*, trans. Burton Watson (New York: Columbia University Press, 1968), p. 142.

[28] Ibid., p. 67.

[29] Ibid., p. 69.

[30] Ibid., p. 281.

[31] Ibid., p. 90.

[32] Ibid., p. 126.

[33] Ibid., p. 131.

of virtue which is closely connected to Dao." Heaven thus models a return to Dao's eternal truth, transcending worldly standards yet anchoring human flourishing.

References

1. Confucius, *The Analects*, trans. D. C. Lau (Penguin Books, 1979).
2. Youlan Feng, *A History of Chinese Philosophy,* trans. Derk Bodde (London, 1952).
3. Ralph Waldo Emerson, *The Portable Emerson*, ed. Jeffery S. Cramer (Penguin Books, 2014).
4. *The Four Books: The Basic Teachings of the Later Confucian Tradition*, trans. Daniel K. Gardner (Indianapolis: Hackett Publishing, 2007).
5. *A Source Book in Chinese philosophy*, trans. Wing-Tsit Chan (Princeton University Press, 1963).
6. Jinhua Jia, 2021. "Writings, Emotions and Obligations: The Religious-Ritual Origin of the Classical Confucian of Cheng (Sincerity)." *Religions* 12 (6). https://doi.org/10.3390/rel12060382
7. Zehou Li, *The Origins of Chinese Thought: From Shamanism to Ritual Regulations and Humaneness*, trans. Robert A. Carleo III, (Brill, 2008).
8. Bo Wang, Re-discussion on "Ran" and "Nature", Philosophical Research, 2018.
9. Zhuangzi, *Zhuangzi: The Essential Writings with Selections from Traditional Commentaries* (Hackett Publishing Company, Inc., 2009).
10. *The Complete Works of Chuang Tzu*, trans. Burton Watson (New York: Columbia University Press).

Chapter 7
Nature: Transcendence Versus Immanence

In this chapter, I mainly explore the essential differences between Emerson and Chinese philosophy in terms of their interpretation of the concept of nature. I argue that Emerson's concept of nature is characterized by its transcendence, while the corresponding concept of nature (Heaven and Earth) in Chinese philosophy is featured by immanence. In terms of transcendence, Roger T. Ames provides us a philosophical definition. He writes, "Strict philosophical or theological transcendence is to assert that an independent and superordinate principle A originates, determines, and sustains B, where the reverse is not the case. Such transcendence renders B absolutely dependent upon A, and thus, nothing in itself."[1] Emerson's metaphysical concept of nature is featured by the strict transcendence. For him, nature is the ultimate cause for all the effects include material nature, institutions, human beings and so on. All of this effects can not be best explained without understanding the metaphysical meaning of nature. In terms of the concept of Heaven and Earth in Chinese philosophy, it doesn't have the meaning of strict transcendence as Roger T. Ames argued. Instead it has the meaning of immanence. On this point, scholars in the field of Chinese philosophy have reached a consensus. Roger T. Ames and David Hall remarks that "We shall continue to argue here, as we have in the past, that one of the most striking features of Chinese intellectual culture from the perspective of the Western interpreter is the absence in any important sense of transcendence in the articulation of its spiritual, moral, and political sensibilities."[2] However, Ame's comment on the Chinese philosophy is only partly true. I argue that the concept of Heaven and Earth in Chinese philosophy has the feature of transcendence. However, it is not strict transcendence as Roger T. Ames has interpreted. It is immanence with a transcendental feature. It signifies transcendence through people's living in harmony with nature. William Franke's critique towards Ame is very insightful. He says,

[1] Roger T. Ames, *Thinking from the Han: Self, Truth, and Transcendence in Chinese and Western Culture* (Albany: State University of New York, 1998), p. 192.

[2] Ibid., p. 193.

S. Gao, *Comparative Studies Between Emerson and Chinese Philosophy*, Nature Aesthetics: Interdisciplinary and Intercultural Dialogue,
https://doi.org/10.1007/978-981-95-0421-3_7

"He stresses that the Way is rather always thoroughly 'interdependent' with human activity and relations. However, while all representations of the Way are certainly always drawn only from human experience, how can this entitle us to deny the possibility of its transcendence of the human?"[3] Although Heaven and Earth is immanent in everyday life of people, it has the transcendental aspect. A life in harmony with nature in Chinese philosophy has the dimension of mysticism. I will examine this point in part III of this book. Below I mainly explain the difference between transcendence and immanence from three pairs of categories: (1) The Spiritual Law versus Heaven and Earth; (2) Life with consciousness of truth, goodness and beauty versus life unfolding in space and time; (3) causality versus substance-function.

1 The Spiritual Law versus Heaven and Earth

The concept of spiritual law is an alternative expression of the concept of nature and God. In Emerson's view, the spiritual law is the metaphysical foundation for the laws of material nature, human society, and human beings' mind. He writes, "The laws of moral nature answer to those of matter as face to face in a glass. 'The visible world and the relation of its parts, is the dial plate of the invisible.' The axioms of physics translate the laws of ethics."[4] According to Emerson, the law of nature is both scientific and moral. At the same time, this law is also a law that human society must follow. Spiritual law has the ontological priority. Emerson opposes guiding human daily moral life with customs and advocates reform. The standard for reform is based on human interpretation of spiritual laws. Although this interpretation will change constantly, Emerson believes that human cognition will evolve with nature and will finally approach the truth.

Emerson's concept of nature creates the dualism between spiritual law and the material world. For him, spiritual law as the creator is superior to the material world. Although Emerson doesn't neglect the importance of material world like Plato, the material world is still treated only as medium for people to connect to the spiritual world. In Chinese philosophy, there is no dualism between the material world and spiritual world. The concept of *qi* is deeper reason why the dualism doesn't exist in the Chinese metaphyscis. In Chinese philosophy, everything is made up of *qi*, and all things are produced by the interaction of two opposing forces of *qi*, namely *yin* and *yang*. The interaction of *yin and yang* is not based on the will of God or any divine consciousness, nor does this interaction have any pre-existing purpose. Instead, *yin* and *yang* interact spontaneously. In the process of interaction of *yin* and *yang*, Dao will manifest itself. In Chinese philosophy, the metaphysical concept of Dao is not transcendental rather it is immanent in the world of human beings. Confucianism

[3] Nahum Brown and William Franke (Eds), *Transcendence, Immanence, and Intercultural Philosophy* (Cham: Palgrave Macmillan, 2016), p. 40.

[4] Ralph Waldo Emerson, *The Portable Emerson*, ed. Jeffrey S. Cramer (Penguin Books, 2014), p. 31.

project the ethical values in the secular world to Dao, which will correspondingly influence the ethical values of the secular world. Xiaogan Liu elucidates this in his interpretation of Chinese philosophy. He points out that Dao in Confucianism doesn't exist in transcendental world which is not concerned with human society. Rather, it is immanent in everyday ethical practices.[5] This stands in stark contrast to Emerson's concept of spiritual law. For him, spiritual law is transcendent and it is the source of truth. It is not based on the experience of everyday life. Instead, it is the foundations for virtue and values in human society. He writes, "For the Maker of all things and all persons stands behind us and casts his dread omniscience through us over things."[6] For Emerson, the values of human beings in society will not have any influence on the ultimate reality. Rather, The Maker of all things and all persons has direct influence on values in the human society. In Chinese philosophy, Dao as the ultimate reality is deeply connected to the values in human society.

In the following, I will explain how the concept of Heaven and Earth is immanent in the world of human beings from two aspects: (1) The morality of Heaven and earth is deeply connected to human ethics; (2) Heaven and Earth is situated through time and space. The Dao is immanent in the lived practices of benevolence, propriety, and relational harmony, binding the cosmic order to the human sphere. Unlike Emerson's transcendent moral force, which humans passively receive from a divine source, Confucian virtue emerges actively within the interplay of interpersonal relationships and the natural world, grounding aesthetics and morality in the tangible rhythms of life.

The concept of Heaven and Earth is deeply connected with morality in this world. This point is well expressed in *The Analects*. "The Master was seriously ill. Tzu-lu asked permission to offer a prayer. The Master said, 'Was such a thing ever done?' Tzu-lu said, "Yes, it was. The prayer offered was as follows: pray thus to the gods above and below.' The Master said, 'In that case, I have long been offering my prayers"[7] Here, Zilu's prayer to the gods reflects a utilitarian motivation—seeking blessings to avoid calamity—while Confucius regards prayer as his lifelong practice of virtue. He does not view Heaven and Earth as deities dispensing favors but as embodiments of ultimate morality, responsive to human ethical conduct. This diverges sharply from Western philosophers' conceptions of ultimate reality. Aristotle's unmoved mover, a rational god, remains aloof from human affairs: "The first mover… is thought thinking itself, and has no concern for the world below."[8] Descartes' God, guarantor of rational order, is equally detached: "God establishes the laws of nature… but does not intervene in human contingencies."[9] Spinoza's

[5] Xiaogan Liu, *Interpretation and Orientation: An Inquiry into the Methodology of Research in Chinese Philosophy* (The Commercial Press, 2009). 刘笑敢:《诠释与定向:中国哲学研究方法之探究》,商务印书馆, 2009.

[6] Ralph Waldo Emerson, *The Portable Emerson*, ed. Jeffrey S. Cramer (Penguin Books, 2014), p. 217.

[7] Confucius, *The Analects*, trans. D. C. Lau (Penguin Books, 1979), p. 91.

[8] Aristotle, *Metaphysics*, trans. W. D. Ross (Oxford: Clarendon Press, 2024), p. 256.

[9] René Descartes, Meditations on First Philosophy, trans. John Cottingham (Cambridge University Press, 1996), p. 54.

God, pure substance, eschews moral attributes: "To attribute will or morality to God is to diminish His perfection."[10] In Confucianism, Heaven and Earth bear ultimate moral value, mirrored in human benevolence and ritual propriety.

Another passage in *The Analects* reinforces this: "When the Master was ill, Zilu had the disciples act as retainers. During a remission, he said, 'This deception of yours has gone on too long! I have no retainers, yet you pretend I do. Whom do I deceive? Heaven itself? I would rather die among my disciples than at the hands of retainers. Even without a grand funeral, will I perish by the roadside?'"[11] Confucian etiquette reserves retainers for the elite, but Confucius, no longer in office, rejects this pretense as a violation of heaven's will. He prioritizes dying authentically among disciples over a ceremonial send-off, upholding benevolence, righteousness, and ritual as heaven's highest values—metaphysically enshrined. Emerson, too, sees ultimate ontology as moral, akin to Plato's Good: "This reality, then, that gives their truth to the objects of knowledge and the power of knowing to the knower, you must say is the idea of good, and you must conceive it as being the cause of knowledge, and of truth in so far as known."[12] Yet Emerson's goodness, flowing from the Oversoul, ensures epistemic truth and cosmic order, not the ethical bonds of Confucianism's human world.

Confucian Heaven and Earth reside within tangible time and space, as seen in *The Analects*: "The Master said, 'What does Heaven ever say? Yet there are the four seasons going round and there are the hundred things coming into being. What does Heaven ever say?'"[13] Heaven lacks language, yet its silent rhythms—the four seasons—produce all things. Time, in Chinese culture, unfolds as life itself, entwined with the land, not a transcendent abstraction. Laozi echoes this in the *Dao De Ching*: "There is a thing confusedly formed, Born before heaven and earth Silent and void. It stands alone and does not change, Goes round and does not weary. It is capable of being the mother of the world. I know not its name So I style it 'the way''. I give it the makeshift name of 'the great'. Being great, it is futher described as receding, Receding, it is described as far away, Being far away, it is described as turning back. Hence, the way is Great; heaven is great; earth is great; and the king is also great. Within the realm there are four things that are great, and the king counts as one."[14] The "realm" in this passage denotes spatiality; Laozi's concept of Dao, Heaven and Earth exist within this concrete expanse, not as transcendental entities.

Thus, Chinese metaphysics remains inseparable from daily life on the land. Understanding this point is crucial for later exploration of the concept of place. Western philosophers often overlook human-land interplay due to their transcendent ontologies. Plato's Forms soar above the material: "The philosopher… prepares for death

---

[10] Baruch Spinoza, *Ethics*, trans. Edwin Curley (Princeton University Press, 1994), p. 96.

[11] Confucius, *The Analects*, trans. D. C. Lau (Penguin Books, 1979), p. 97.

[12] Plato, *The Republic* (Vol II: Books 6-10), trans. Paul Shorey (Harvard University press, pp. 103–105).

[13] Confucius, *The Analects*, trans. D. C. Lau (Penguin Books, 1979), p. 146.

[14] Laozi, *Tao Te Ching*, trans. D. C. Lau (The Chinese University Press, 1963), p. 37.

to ascend to the eternal."[15] In contrast, Confucianism and Daoism denies such transcendence. Heaven and Earth is the process of unfolding life in specific time and space. Emerson's concept of life, however, is a spiritual essence beyond matter: "For this communication is an influx of the Divine mind into our mind. It is an ebb of the individual rivulet before the flowing surges of the sea of life."[16] Life in this passage refers to universal life of Oversoul or God. It is not connected to the physical land. Bodily existence, unbound by locality, reflects the Oversoul's creation. Emerson values life more than valuing physical life. Although he has an aesthetic appreciation of the physical nature, which is in place and time, he stresses that the beauty of physical nature is not the final end. It is our medium to know God. "But beauty in nature is not ultimate. It is the herald of inward and eternal beauty, and is not alone a solid and satisfactory good. It must stand as a part, and not as yet the last or highest expression of the final cause of Nature."[17] For Emerson, the beauty of nature is the result of spiritual law. If we attach to the beauty of nature for its own sake, or we attach to the land, we will be separated from the ontological reality, which is the spiritual law.

2 Life Transcending Time and Space Versus and Life Unfolding in Time and Space

The concept of life is an important term in Emerson and Chinese philosophy. It can help us explain further the difference between transcendence and immanence in Emerson and Chinese philosophy. Emerson's concept of nature is in essence the spiritual life which transcends time and space. In contrast, the concept of life in Chinese philosophy is deeply connected with time and space. The concept of life in Emerson is intimately connected to nature and God, while in Chinese philosophy, life is tied to Heaven and Earth, *qi*, and Dao.

In terms of the logical relationship among life, nature, and God in Emerson, the following two passages give us a good insight. "Embosomed for a season in nature, whose floods of life stream around and through us, and invite us, by the powers they supply, to action proportioned to nature, why should we grope among the dry bones of the past, or put the living generation into masquerade out of its faded wardrobe?"[18] "I have been led yesterday into a rambling exculpatory talk on theism. I say that here we feel at once that we have no language; that words are only auxiliary and not adequate, are suggestions and not copies of our cognition. I deny personality to God because it is too little, not too much. Life, personal life, is faint and cold to the energy

---

[15] Plato, *The Republic* (Vol II: Books 6-10), trans. Paul Shorey, Loeb Classical Library (Cambridge, MA/London: Harvard University Press, 1935), pp. 103–105.

[16] Ralph Waldo Emerson, *The Portable Emerson*, ed. Jeffrey S. Cramer (Penguin Books, 2014), p. 217.

[17] Ibid., p. 27.

[18] Ibid., p. 17.

of God. For Reason and Love and Beauty, or that which is all these, is the life of life, the reason of reason, the love of love."[19] From the above two passages, it can be seen that Emerson's concept of life is transcendental for the following reasons: (1) God represents life, which is not personal but impersonal. It presents ultimate values of truth, goodness, and beauty; (2) Life transcends time and space, and it is spiritual rather than material.

In contrast, the concept of life in Chinese philosophy is grounded in the concept of *qi*. Life emerges through the interaction of two opposing and complementary forces of *yin* and *yang*. The interaction of *yin* and *yang* are guided by values. Confucianism stresses moral values while Daoism focuses on transcendence of moral values, which is termed as self-so. "The way begets one; one begets two; two begets three; three begets the myriad creatures."[20] *qi* and Dao intertwine, with *qi* as the unseen constituent of nature and humanity: "The seminal kernel of vitality and the spirit are born from the Course. It is from this seminal kernel that the physical body is originally born."[21] Dao resides within qi's transformations, its spirituality inseparable from yin and yang's dialectic. Scholar Robin Wang generalizes the different relationships of yin and yang into six forms: (1) Contradiction and opposition; (2) Interdependence; (3) Mutual inclusion; (4) Interaction or resonance; (5) Complementarity or mutual support; (6) Change and transformation."[22] This dynamic births life inevitably, distinct from Emerson's divine perfection. For Emerson, life flows from God's spiritual law, externalized in matter as a symbol of truth, goodness, and beauty. In Chinese thought, *yin* and *yang*'s relational spirit—neither law nor perfection—engenders life's richness.

The concept of Heaven and Earth in Confucianism and Daoism differs significantly from Emerson's understanding of God, which he views as a transcendental entity existing beyond time and space. In contrast, the Chinese philosophical notion of Heaven and Earth is immanent, situated within the dimensions of time and space, and closely tied to the cycle of seasons. In Chinese thought, seasons are intricately linked to agricultural production, serving as the foundation for generating and sustaining life. This marks a fundamental distinction between Emerson's philosophy and Chinese philosophical traditions. To illustrate this, I will cite two passages from Zhuangzi. "The four seasons arrive one after the other, and the ten thousand things are produced accordingly—now flourishing, now declining, now peaceful, now warlike, all grouping into their own regularities. Sometimes clear, sometimes turbid, yin and yang adjust to each other, forming a harmony."[23] "When the energy

[19] Ralph Waldo Emerson, *Journals of Ralph Waldo Emerson* (1820-1872 Vol IV), ed. Edward Waldo Emerson and Waldo Emerson Forbes (Boston and New York: Houghton Mifflin Company), pp. 471–472.

[20] Laozi, *Tao Te Ching*, trans. D. C. Lau (The Chinese University Press, 1963), p. 63.

[21] Zhuangzi, *Zhuangzi: The Essential Writings with Selections from Traditional Commentaries*, trans. *Brook Ziporyn* (Indianapolis: Hackett Publishing, 2009), p. 87.

[22] Robin R. Wang, *YINYANG: The Way of Heaven and Earth in Chinese Thought and Culture* (Cambridge University Press, 2012), pp. 8–12.

[23] Zhuangzi, *Zhuangzi: The Essential Writings with Selections from Traditional Commentaries*, trans. Brook Ziporyn (Hackett Publishing Company, 2009), p. 67.

of spring bursts forth, all the plants come to life. When the treasures of the earth get full access to the autumn, they all mature for harvest. And do you think spring and autumn can do this without availing themselves of something? It is the Course of Heaven that is already 'moving in them!'"[24] For Zhuangzi, the seasons are manifestation of *yin* and *yang* energies, embodying the cyclical interplay of opposites that generate and sustain life. Dao is immanent in the process of production of life through energies of yin and yang throughout different seasons. By contrast, for Ralph Waldo Emerson, God exists as a transcendent entity, wholly beyond time and space. When God creates life in Emerson's philosophy, this act is grounded in eternal spiritual principles—such as the divine intellect or the Oversoul—rather than any material processes or conditions. For Emerson, the material world serves merely as a reflection or symbol of this higher spiritual reality. In Chinese philosophy, both Confucianism and Daoism share the same vision that life is the result of interaction of yin and yang energy which are operative in the physical world. Dao is immanent in the process of generation of life. However, there is still transcendental aspect of Dao, which has direct influence on all aspects of life. I will discuss it in the following chapters.

3 Causality Versus Substance-Function

Emerson's concept of causality and the idea of substance-function in Chinese philosophy are two terms that can help us grasp Emerson's transcendence and immanence in Chinese philosophy. Emerson's concept of nature is grounded in causality, contrasting starkly with the substance-function paradigm of Heaven and Earth in Chinese philosophy. For Emerson, causality implies an ultimate cause behind all existence, inexorably producing the material world as its effect—nature as the fruit of divine creation. In Chinese philosophy, however, Heaven and Earth lack this generative agency; their relationship with all things is not causal but substantive-functional. Heaven and earth form the substance, manifesting through the operations of myriad entities, which arise as condensations of *qi*—an eternal, indestructible essence. Let us first examine Emerson's causal framework, then juxtapose it with Kant's causality and the Chinese model.

Emerson's concept of causality is directly tied to his two concepts of nature: *natura naturans* (nature creating) and *natura naturata* (nature created). He writes, "There is a dynamic nature—natura naturans—and a nature resultant—natura naturata. The world proceeds from the same spirit as the body of man. It is a remoter and inferior incarnation of God, a projection of God in the unconscious."[25] For him, natura naturans, the dynamic force, acts as the ultimate cause, yielding natura naturata, the material outcome. For him, we are born believers in causality. This causality

[24] Ibid., p. 95.

[25] Emerson, *The Collected Works of Ralph Waldo Emerson*, Voume I: Nature, Addresses, and Lectures, ed. Robert E. Siller and Alfred R. Ferguson (Cambridge, MA: Harvard University Press, 1971), pp. 35–36.

inverts Kant's framework. Kant bifurcates reality into phenomena (the sensory world we perceive) and noumena (things-in-themselves, unknowable in essence): "Buty even if we could say anything synthetically **about things in themselves** through the pure understanding (which is nevertheless impossible), this still could not be related to appearances at all, which do not represent things in themselves."[26] For Kant, pure understanding such as causality cannot be used to understand the things in themselves. We can never know the things in themselves. What we can know is appearance. Emerson, however, thinks that God as the ultimate reality can be known through interpretation of nature. "Nature is the incarnation of a thought, and turns to a thought again, as ice becomes water and gas."[27] This nature embodies moral goodness and beauty, unlike Kant's view, where goodness resides in human free will: "*Will* is a kind of causality of living beings insofar as they are rational, and *freedom* would be that property of such causality that it can be efficient independently of alien causes *determining* it, just as *natural necessity* is the property of the causality of all nonrational beings to be determined to activity by the influence of alien causes."[28] Kant's free will acts autonomously, unbound by external forces like God, whereas Emerson locates true freedom in surrendering to divine causality: "…and the heart which abandons itself to the Supreme Mind finds itself related to all its works, and will travel a royal road to particular knowledges and powers."[29]

In Chinese philosophy, the concept of Heaven and Earth can not be understood in terms of Emerson's concept of causality. For Emerson, *natura naturans* as an efficient nature is a creator, and *natura naturata* is the effect of the creation. There is dualism between cause and effect. Cause is superior to effect since it is pure spiritual. In contrast, the concept of Heaven and Earth can be best explained through substance-function. In Chinese philosophy, there is no dualism. All things in nature are eternal, and their existence is not caused by Heaven and Earth. Rather, the spontaneous transformations of *qi* explains the generation of life. Qi is both material and spiritual and qi is not created. In this aspect, both Confucianism and Daoism share the same vision. "Heaven is my father and Earth is my mother, and even such a small creature as I finds an intimate place in their midst."[30] For Zhangzai, Heaven and Earth gives human beings life through the interaction of *yin* and *yang*. Zhuangzi has the same view on it. He states, "Heaven and earth are born together with me, and the ten thousand things and I are one."[31] In Chinese philosophy, all things are the result of

[26] Immanuel Kant, *Critique of Pure Reason*, trans. Paul Guyer and Allen W. Wood (Cambridge University Press, 1998), p. 375.

[27] Ralph Waldo Emerson, *The Best Read Naturalist*, ed. Michael P. Branch and Clinton Mohs (University of Virginia Press, 2017), p. 163.

[28] Immanuel Kant, *Groundwork of the Metaphysics of Morals*, trans. Mary Gregor (Cambridge University Press, 1998), p. 52.

[29] Ralph Waldo Emerson, *The Portable Emerson*, ed. Jeffrey S. Cramer (Penguin Books, 2014), p. 215.

[30] *A Source Book in Chinese Philosophy*, trans. Wing-tsit Chan (Princeton University Press, 1963), p. 497.

[31] Zhuangzi, *Zhuangzi: The Essential Writings with Selections from Traditional Commentaries, trans. Brook Ziporyn* (Hackett Publishing Company, 2009), p. 15.

transformation of *qi*. There is no justification for the existence of Heaven and Earth in Chinese philosophy. The ontological status of *qi* is taken for granted. There is no dualism between Heaven & Earth and all things in nature in Chinese philosophy. They are united and exist within space and time.

References

1. Roger T. Ames, *Thinking from the Han: Self, Truth, and Transcendence in Chinese and Western Culture* (Albany: State University of New York, 1998).
2. *Transcendence, Immanence, and Intercultural Philosophy*, ed. Nahum Brown and William Franke (Cham: Palgrave Macmillan, 2016).
3. Ralph Waldo Emerson, *The Portable Emerson*, ed. Jeffrey S. Cramer (Penguin Books, 2014).
4. Xiaogan Liu, *Interpretation and Orientation: An Inquiry into the Methodology of Research in Chinese Philosophy* (The Commercial Press, 2009). 刘笑敢:《诠释与定向:中国哲学研究方法探究》,商务印书馆, 2009.
5. Confucius, *The Analects*, trans. D. C. Lau (Penguin Books, 1979).
6. Aristotle, *Metaphysics*, trans. W. D. Ross (Oxford: Clarendon Press, 1924).
7. René Descartes, *Meditations on First Philosophy*, trans. John Cottingham (Cambridge University Press, 1996).
8. Baruch Spinoza, *Ethics*, trans. Edwin Curley (Princeton University Press, 1994).
9. Plato, *The Republic* in *Plato in Twelves Volumes*, trans. Paul Shorey (Harvard University Press).
10. Laozi, *Tao Te Ching*, trans. D. C. Lau (The Chinese University Press, 1963).
11. Plato, *Phaedo*, trans. G. M. A. Grube (Indianapolis: Hackett, 1977).
12. Emerson, *The Collected Works of Ralph Waldo Emerson*, Volume I : Nature, Addresses, and Lectures, ed. Rober E. Spiller and Alfred R. Ferguson (Cambridge, MA: Harvard University Press, 1971).
13. Ralph Waldo Emerson, *Journals of Ralph Waldo Emerson*, 1838–1842, ed. Edward Waldo Emerson and Waldo Emerson Forbes (Boston: Houghton Mifflin, 1911).
14. Zhuangzi, *Zhuangzi: The Essential Writings*, trans. Brook Ziporyn (Indianapolis: Hackett Publishing, 2009).
15. Robin Wang, *Yinyang: The Way of Heaven and Earth in Chinese Thought and Culture* (Cambridge University Press, 2012).
16. Immanuel Kant, *Critique of Pure Reason*, trans. Paul Guyer and Allen W. Wood (Cambridge University Press, 1998).
17. Immanuel Kant, *Groundwork of the Metaphysics of Morals*, trans. Mary Gregor (Cambridge University Press, 1997).
18. *A Source Book in Chinese Philosophy*, ed. Wing-tsit Chan (Princeton University Press, 1963).

Part III
Self-Transcendence and Virtue

In part II of the book, I explore the metaphysics of Ralph Waldo Emerson and Chinese philosophy, specifically the interplay between nature and ultimate value. In this part, I will shift focus to examine the relationship between self and virtue. However, to lay the groundwork for understanding this relationship, it is first necessary to briefly address the connection between metaphysics and virtue. Both Emerson and Chinese philosophy share a fundamental conviction: living a good life is the paramount concern. For them, metaphysics serves ethics. Emerson explicitly conveyed this point in his journal, "Transcendentalism, Metaphysics and Ethics look inwards."[1] Here, the comprehension of metaphysics acts as a ladder for human moral progress. In this regard, Emerson aligns with Chinese philosophy, as both assert that a person's most authentic self is embodied in virtue. For Emerson, the central concept expressing the self is the "soul," with "self-reliance" as its core virtue. In Chinese philosophy, particularly Confucianism and Daoism, the self is articulated through the concept of "mind" (心, xin). For Confucianism, the virtues of the mind are benevolence (仁, ren) and righteousness (义, yi); for Daoism, it is the state of "free and easy wandering" (逍遥游, xiaoyao you).

---

[1] Ralph Waldo Emerson, *Journals of Ralph Waldo Emerson*, Houghton Mifflin Company, 1842, p. 73.

Chapter 8
Self: The True Self

In this chapter, I will discuss the concept of the true self and the false self. For both Emerson and Chinese philosophy, there are two types of self: the true self and the false self. The true self is closely related to virtue. In terms of the idea of true self and false self, they use different concepts to express two kinds of self. Emerson often employs terms such as "real self", "the inner self", "aboriginal self" to express the idea of the true self. In terms of the false self, he uses terms such as "outer self", "egotism", "self-willed", "individualism" and "prison" to express it. In Chinese philosophy, the self is often expressed through the following words, wu (吾),ziji (自己), and wo (我). In these words, they include both the idea of the true self and the false self. In both Emerson and Chinese philosophy, the essence of true self is virtue, though they interpret it differently. In the following, I will explain the core category of soul and mind and discuss their relationship with the true self and virtue in the works of Emerson and Chinese philosophy.

1 The true self in Emerson: Soul and Virtue

During Emerson's lifetime, the concept of self carried negative connotations. As Emerson scholar Wesley T. Mott observes, "As a young minister at the Second Church in Boston (1829-1832), Emerson was already wrestling with the implications of the 'self'—an unfashionable term in the late twentieth century, when literary theorists have treated 'self-construction,' according to Daniel Walker Howe, 'as essentially pathological.'"[1] However, Emerson reconstructs the concept of self through the virtue of self-reliance. For him, the true self is related to soul and virtue. The interrelationship among the true self, soul and virtue is best expressed in his essay "The Oversoul". He writes, "What we commonly call man, the eating, drinking, planting, counting

[1] Joel Myerson, *Historical Guide to Ralph Waldo Emerson* (Oxford University Press, 2018), p. 74.

S. Gao, *Comparative Studies Between Emerson and Chinese Philosophy*, Nature Aesthetics: Interdisciplinary and Intercultural Dialogue,
https://doi.org/10.1007/978-981-95-0421-3_8

man, does not, as we know him, represent himself, but misrepresents himself. Him we do not respect, but the soul, whose organ he is, would he let it appear through his action, would make our knees bend. When it breathes through his intellect, it is genius; when it breathes through his will, it is virtue; when it flows through his affection, it is love. And the blindness of the intellect begins when it would be something of itself. The weakness of the will begins when the individual would be something of himself. All reform aims in some one particular to let the soul have its way through us; in other words, to engage us to obey."[2] For Emerson, the soul represents the true self of a person. In the history of Western philosophy, the concept of the soul has been employed to represent the self since ancient Greek times. Aristotle, for instance, explored the soul in his *Nicomachean Ethics*, though his interpretation diverges from Emerson's.

Aristotle categorizes the soul into three types: the vegetative soul, responsible for nutrition and bodily growth; the appetitive soul, tied to desire; and the rational soul, which he considers the essence of human identity. For Aristotle, the vegetative and appetitive aspects of the soul do not define the essence of a person whereas the rational soul does. Emerson, however, rejects such divisions. He views the soul as transcendental which has the power to ignite mental faculties of intellect, memory, and imagination. However, he stresses that the soul is not a physical organ. In the above quote, Emerson asserts that when the soul "breathes through his will", it manifests as virtue. Emerson's conception of will thus contrasts sharply with Immanuel Kant's notion of the "good will". Kant writes, the good will autonomously establishes and adheres to moral laws. For Kant, every human being has a will which can legislate universally through all its maxims".[3] Emerson, however, envisions the subordination of self-will to the soul's animating influence, through which virtue attains its fullest realization. For him, self-will represents the false conception of self which should be transcended. The foundational virtue of Emerson's ethical framework is "self-reliance", a compound that fuses selfhood with relational dependence. Contrary to some interpretations, self-reliance is not the autonomy of an isolated individual. Wesley T. Mott in his "The Age of the First Person Singular" mentions Quentin Anderson's criticism towards Emerson's idea of self. "Quentin Anderson memorably charged Emerson with constructing an 'imperial self' that results in 'a denial of history, membership in a generation, charity, reform, institutional means of every sort.'"[4] Anderson overlooks the metaphysical dimension of Emerson's concept of self. His idea of self is based on reliance on God. From the metaphysical aspect, all things are interrelated. Emerson writes, "We live in succession, in division, in parts, in particles. Meantime within man is the soul of the whole; the wise silence; the universal beauty, to which every part and particle is equally related; the eternal

[2] Ralph Waldo Emerson, *The Portable Emerson*, ed. Jeffrey S. Cramer (Penguin Books, 2014), p. 212.

[3] Immanuel Kant, *Groundwork of the Metaphysics of Morals*, trans. Allen W. Wood (Yale University Press, 2018), p. 44.

[4] Joel Myerson, *Historical Guide to Ralph Waldo Emerson* (Oxford University Press, 2000), p. 68.

ONE."[5] Thus, self-reliance reflects a profound connection between humanity and the divine, a relationship that Emerson sees as essential to human existence. Emerson's self-reliance is based on on self-trust. Without it, self-reliance cannot stand. He clarifies, "Self-Trust, that is, not a faith in man's own whim or conceit as if he were quite severed from all other beings and acted on his own private account, but a perception that the mind common to the Universe is disclosed to the individual through his own nature."[6] From his interpretation of self-trust, we can find that his understanding of self is relation-oriented. Self-trust is the belief in the existence of universal mind, which will be disclosed to every person's own nature. Without the virtue of self-trust, self-reliance is impossible.

2 The True Self in Confucianism: Mind and Virtue

The Chinese expression for "self" is 自己. It is made up of two Chinese characters: zi (自)and ji (己). According to the old Chinese dictionary, *shuowen jiezi*, zi (自)refers to the nose. When people converse with each other, they tend to point to their nose to refer to themselves. Later, it was extended to denote "self". The old Chinese dictionary Shuowen Jiezi explains "ji" as pictographic text. The shape of oracle bone script, like the shape of a twisted rope. Shen Xu pointed out that "ji" is the ancient word "ji", borrowed for "self" use. The Chinese word 纪(ji) means discipline or restraint. From this entymology, we can see that ji 己 has close relationship with virtue. The core terms for virtues that Confucianism stresses are benevolence and righteousness. The following passage from *Mencius* best addresses the relationship among self, mind and virtue.

> Hence it is said: all palates have the same preference in taste; all ears in sound; all eyes in beauty. Should hearts prove to be an exception by possessing nothing in common? What is it, then, that is common to all hearts? Reason and rightness. The sage is simply the man first to discover this common element in my heart. Thus reason and rightness please my heart in the same way as meat pleases my palate.[7]

When I say that all men have the mind which cannot bear to see the suffering of others, my meaning may be illustrated thus: Now, when men suddenly see a child about to fall into a well, they all have a feeling of alarm and distress, not to gain friendship with the child's parents, nor to seek the praise of their neighbors and friends, nor because they dislike the reputation [of lack of humanity if they did not rescue the child]. From such a case, we see that a man without the feeling of commiseration is not a man; a man without the feeling of shame and dislike is not a man; a man without the feeling of deference and compliance is not a man; The

[5] Ralph Waldo Emerson, *The Portable Emerson*, ed. Jeffrey S. Cramer (Penguin Books, 2014), p. 211.

[6] The Early Lectures of Ralph Waldo Emerson, ed. Robert E. Spiller, and Wallace E. Williams. Volume 2. Cambridge, MA: Harvard University Press, 1959-1972), p. 151.

[7] Mencius, *Mencius*, trans. D. C. Lau (Penguin Books, 1970), p. 164.

feeling of commiseration is the beginning of humanity; the feeling of shame and dislike is the beginning of righteousness; the feeling of deference and compliance is the beginning of propriety; and the feeling of right and wrong is the beginning of wisdom.[8]

The translation of the mind in this passage tends to be misleading. The Chinese character for the word of mind is 心(xin) and it is often combined with the concept of *qi* (vital energy) and *qing* (emotion) in everyday language. From this usage, we can find the Chinese concept of mind is deeply related to body and feelings. I will explain the mind from the following three aspects. (1) Mind is the bearer of moral virtues especially humanity and righteousness. The justification of the moral virtues in mind is based on an analogy. For Mencius, human beings have common aesthetic experience of taste, sound, and flavor and therefore our mind will have common sense of moral values. This justification is not based on rational judgement as Kant has done in his *Metaphysics of Morals*. For him, morals are not based on experiences but on metaphysical principles. (2) Mencius' concept of mind is related to intersubjective feelings which are generated through deep interaction with people within close relationships. They are not like the feeling of wonder as Aristotle has mentioned in the beginning of his *Metaphysics*. Hear-mind are the feelings towards the people. Intersubjective feelings for Mencius are the foundation of four cardinal virtues which include benevolence, righteousness, propriety and wisdom. Intersubjective feelings are generated by emotion-oriented reason. (3) Mencius stresses cultivation of feelings among the people and develop those feelings to become stronger and deeper. He also stresses the extension of these feelings to other people. Without stronger feelings towards people, we will not fulfil our obligations towards them.

The core virtue in Confucianism is benevolence. The virtue of benevolence is paramount. All other virtues revolve around benevolence. For example, propriety is considered as important virtue. However, the purpose of propriety is to secure the virtue of benevolence. Without benevolence, propriety will be considered as useless. The virtue of wisdom is also important and it is based on benevolence. Wisdom is not about the knowledge of natural world or the world relating to higher powers such as God, rather it is related to rich knowledge of the human world such as human society, rituals and so on. All of the knowledge will strengthen the harmony among the people.

3 The True Self in Zhuangzi: Empty Mind and Virtue

Zhuangzi's idea of the true self is developed from the reflection and criticism of Confucius virtue. Compared with Confucianism, which stresses mind, Zhuangzi stresses the concept of empty mind. For him, an empty mind is closely related to

[8] *A Source Book in Chinese Philosophy*, trans. Wing-Tist Chan (Princeton University Press, 1963), p. 65.

virtue. He says, "So to subordinate your inborn nature to Humanity and Responsibility, even if you succeed like Zeng and Shi, is not what I call good. To subordinate your inborn nature to the Five Flavors, even if you succeed like Yu Er, is not what I call good. To subordinate your inborn nature to the Five Tones, even if you succeed like Master Kuang, is not what I call acute hearing. To subordinate your inborn nature to the Five Colors, even if you succeed like Li Zhu, is not what I call acute vision. What I call good is not Humanity and Responsibility, but just being good at your own Virtuosity."[9] It can be seen that Zhuangzi's understanding of virtue is also a kind of goodness. However, Confucianism and Daoism differ in their interpretation of the idea of goodness. For Confucianism, the virtue of benevolence and righteousness is goodness while for Daoism, the goodness lies in the full development of human nature. Regarding human nature, Zhuangzi uses the category of mind to express it. But unlike Confucianism, Zhuangzi emphasizes the mind as "empty mind" or "constant mind". When our mind is in the state of wandering, human nature is fully realized. When the human mind achieves the state of wandering, the power of human life will be fully exerted, thus realizing Zhuangzi's virtue. This thought is well explained in the following paragraph. "The consciousness of living creatures depends on their breath. It is not Heaven's fault if it becomes depleted. Heaven blows through them day and night without cease, but human beings see to it that all their openings are blocked off. Only when you reopen yourself all the way back down to the placenta can your mind wander in the Heavenly. When there is insufficient space in a house, the womenfolk will surely fall to quarreling. If the mind has no Heavenly wandering, the six apertures of perception will interfere with one another. This is why being in a vast mountain forest benefits a man more than whatever might be granted by the gods."[10]Zhuangzi's concept of wandering is a virtue term. It doesn't mean the physical roaming from one place to another, rather it is a reflective state of mind I will explore further in part III. Wandering mind for Zhuangzi represents the core virtue of human beings. Then what is heavenly wandering? Zhuangzi never defines the term. I will address this concept from the following aspects. (1) It is related to our mind. For Zhuangzi, when our mind is empty of the false self and full of the true self., mind will achieve heavenly wandering; (2) It is also related to our body since there is no separation between mind and body in Zhuangzi. The body will be full of vital energy if our mind is full of the true self; (3) Heavenly wandering is the result of rational reflection.

References

1. Aristotle, *Nicomachean* (Oxford University Press, 2009).
2. Ralph Waldo Emerson, *The Portable Emerson*, ed. Jeffrey S. Cramer (Penguin Books, 2014).

[9] *A Source Book in Chinese Philosophy*, trans. Wing-Tsit Chan (Princeton University Press, 1963), p. 65.

[10] *Zhuangzi, Zhuangzi: The Essential Writings with Selections from Traditional Commentaries*, (Indianapolis: Hackett Publishing Company, 2009), p. 113.

3. Immanuel Kant, *Groundwork of the Metaphysics of Morals*, trans. Allen W. Wood (Yale University press, 2018).
4. Joel Myerson, *Historical Guide to Ralph Waldo Emerson* (Oxford University Press, 2000).
5. *The Early Lectures of Ralph Waldo Emerson*, ed. Robert E. Spiller and Wallace E. Williams, vol. 2 (Cambridge, MA: Harvard University Press, 1959-1972).
6. Mencius, *Mencius*, trans. D. C. Lau (Penguin Books, 2004).
7. *A Source Book in Chinese Philosophy*, trans. Wing-Tist Chan (Princeton University press, 1963).
8. Zhuangzi, *Zhuangzi: The Essential Writings with Selections from Traditional Commentaries*, trans. Brook Ziporyn (Indianapolis/Cambridge: Hackett Publishing Company, 2009).

Chapter 9
Virtue: Power and Values

The concept of virtue in Emerson and Chinese philosophy is closely related to power and value. The concepts of power and value are both metaphysical concepts and ethical concepts. Metaphysical meanings of power and value as I examined in part I serve as the foundation for ethical concepts of power and value. The foundational value is goodness, which is deeply related to truth and beauty although Emerson and Chinese philosophy have different interpretation of these values. In this chapter, I mainly explain the concept of virtue in Emerson and Chinese philosophy from its relationship with power and value: (1) The power and value of cognition; (2) The power and value of the body; (3) The power and value of action.

1 Virtue: Cognitive Power and Values

In Emerson and Chinese philosophy, virtue is closely related to cognitive abilities. When they define cognitive abilities, they often use the concepts of power and value to explain them. For Emerson, these cognitive abilities include reason, understanding, intellect, and imagination. For him, these abilities are powers which coming directly from God. Power relates to the ultimate values of truth, goodness, and beauty. Confucianism emphasizes emotional rationality, while Daoism focuses on reflective rationality. The power brought by emotional rationality and reflective rationality is closely related to value. The logical relationship between cognitive powers and values is as follows: cognitive powers are directed by ultimate values which are the combination of goodness, truth and beauty. Cognitive power will cease to exist without these values.

S. Gao, *Comparative Studies Between Emerson and Chinese Philosophy*, Nature Aesthetics: Interdisciplinary and Intercultural Dialogue,
https://doi.org/10.1007/978-981-95-0421-3_9

1.1 Virtue in Emerson: Cognitive Abilities, Power and Values

Robert D. Richardson is the biographer of Emerson, and he made this evaluation of Emerson's life, "Emerson lived for ideas, but he did so with the reckless, headlong ardor of a lover. He associated the human mind and its capacity for thought with activity and energy. He hated the passive notion of the mind as a blank slate. He concentrated instead on the individual's sources of power, on access to the central fires that ignite the mind."[1] Emerson's power is interrelated to cognitive abilities and value. Cognitive faculties occupy a major position in Emerson's understanding of self. He states, "Knowing is the measure of the man. By how much we know, so much we are. It is the measure of the man, because it is the measure of God."[2] Emerson views cognition as a measure of a person, representing the virtuous self. Human cognitive abilities include reason, understanding, intellect, and imagination. When expressing these cognitive abilities, Emerson often uses the word "power", which is closely intertwined with the ultimate values he emphasizes: truth, goodness, and beauty.

First, reason and understanding constitute interrelated yet distinct cognitive abilities. Reason is superior to understanding. What corresponds to understanding is experiential knowledge, which is analyzable. Reason, on the other hand, is holistic and intuitive, a revelation of the cosmic mind. In his 1835 essay titled "On the Nature of the mind", Emerson explains these two different cognitive abilities as follows: "Every property of matter is a school for the understanding,—its solidity or resistance, its inertia, its extension, its figure, its divisibility. The understanding adds, divides, combines, measures, and finds nutriment and room for its activity in this worthy scene. Meantime, Reason transfers all these lessons into its own world of thought, by perceiving the analogy that marries Matter and Mind."[3] According to Emerson, the characteristics of reason are eternity and intuition which are, revelations from God, and through rationality, one can directly know God. Understanding is an executive ability, operating between the mind and inert matter, working in time and space, and thus is continuous. It divides, compares, reasons, and invents. It lives from reason, yet often disobeys it. It controls the material world, but often for the pleasure of the senses. Emerson's emphasis on the superiority of reason is also reflected in the following text: "To the senses and the unrenewed understanding, belongs a sort of instinctive belief in the absolute existence of nature. In their view man and nature are indissolubly jointed. Things are ultimates, and they never look beyond their sphere. The presence of Reason mars this faith. The first effort of thought tends to relax this despotism of the senses which binds us to nature as if we were a part of it, and shows us nature aloof, and, as it were, afloat. Until this higher agency intervened, the animal eye sees, with wonderful accuracy, sharp outlines and colored surfaces. When the

---

[1] Robert D. Richard Jr., *Emerson: The Mind on Fire* (University of California Press, 1995), xi.

[2] Ralph Waldo Emerson, *The Best Read Naturalist*, ed. Michael P. Branch and Clinton Mohs (University of Virginia Press, 2017), p. 209.

[3] Ralph Waldo Emerson, *The Portable Emerson*, ed. Jeffrey S. Cramer (Penguin Books, 1946), p. 33.

eye of Reason opens, to outline and surface are at once added grace and expression. These proceed from imagination and affection, and abate somewhat of the angular distinctness of objects. If the Reason be stimulated to more earnest vision, outlines and surfaces become transparent, and are no longer seen; causes and spirits are seen through them. The best moments of life are these delicious awakenings of the higher powers, and the reverential withdrawing of nature before its God."[4] In this passage, Emerson uses the concept of power to explain human reason. Understanding focuses only on the facts of matter itself, viewing the facts of matter as the ultimate existence. It sees humans as part of nature, with an unbreakable connection between humans and nature. But reason elevates human cognition, starting from imagination and emotion, seeing the reasons and spirits behind these materials, and in the face of God, humans begin to withdraw from nature. This ordering of cognitive faculties is crucial to our understanding of Emerson's aesthetic appreciation of nature (which I will elaborate on in the last chapter of the book).

Regarding Emerson's rationality, it is important to emphasize that the reason he speaks of is not human reason. In his view, the human mind only has receptivity and does not have the ability to generate concepts on its own. Reason is the insight received from the highest entity, God. Comparing Emerson's reason with Kant's rationality can better understand this point. Kant's concept of pure reason does not correspond to the outside natural world. He writes, "There is of reason, as there is of the understanding, a purely formal, that is, logical use, in which reason abstracts from all contents of knowledge. But there is also a real use, insofar as reason itself contains the origin of certain concepts and principles which it has not borrowed either from the senses or from the understanding."[5] Kant's concept of reason does not connect with the intuitive objects obtained by sensory cognition, but transcends beyond specific experiences, grasping unconditional absolute totality. The form of rational apprehension is the idea of reason, which is a necessary concept of pure reason for which no corresponding object can be found in sensory experience. The idea is proposed by human pure reason. It is issued by human reason itself, and it has nothing to do with the ultimate facts of the world or the existing objects of the material world.

Emerson regards human rational cognition as the foundation of the value of human beings. However, in the real world, people often measure the value of a person based on wealth, status, or the professions that society values at the time, but Emerson criticizes this. He does not deny these worldly social aspects such as family, profession, or social status, believing that these are the foundations for individuals to realize themselves. He writes in his essay titled Culture, "We must have an intellectual quality in all property and in all action, or they are naught. I must have children, I must have events, I must have a social state and history, or my thinking and speaking want body or basis. But to give these accessories any value, I must know them as contingent and rather showy possessions, which pass for more to the people than to me. We see

[4] Ibid., p. 40.

[5] Immanuel Kant, *Critique of Pure Reason*, trans. Marcus Weigelt (Cambridge University Press, 2007), pp. 288-289.

this abstraction in scholars, as a matter of course; but what a charm it adds when observed in practical man."[6] According to Emerson, everyone has reason. It is the channel for God to enter the human soul. Only when human rationality is opened, can people truly realize the virtue of self-reliance.

Second, intellect is another cognitive ability Emerson stresses in his work. He defines intellect by using the concept of power. He writes, "Intellect is the simple power anterior to all action or construction."[7] For Emerson, cognitive power is not produced through human actions, or practices, but is a simple power coming from God. God for Emerson is not a person rather is full of consciousness of truth, goodness, and beauty. He gives an account of intellect in his essay titled Intellect, "Intellect is void of affection, and sees an object as it stands in the light of science, cool and disengaged. The intellect goes out of the individual, floats over its own personality, and regards it as a fact, and not as *I* and *Mine*. He who is immersed in what concerns person or place cannot see the problem of existence. This the intellect always ponders. Nature shows all things formed and bound. The intellect pierces the form, overleaps the wall, detects intrinsic likeness between remote things, and reduces all things into a few principles."[8] The intellect that Emerson stresses is devoid of emotion. The object of intellect is the cognition of nature. He emphasizes neutrality of cognition since it is from the perspective of science. So, what is the relationship between intellect and virtue? Emerson writes, "If we consider what persons have stimulated and profited us, we shall perceive the superiority of the spontaneous or intuitive principle over the arithmetical or logical. The first contains the second, but virtual and latent. We want, in every man, a long logic; we cannot pardon the absence of it, but it must not be spoken. Logic is the procession or proportionate unfolding of the intuition; but its virtue is as silent method; the moment it would appear as propositions, and have a separate value, it is worthless."[9] Intuition in this passage refers to intellect. For him, intellect is superior to logic. Logic in our mind has virtue. But the virtue in the intellect is higher than virtue in the ability to do the logic. Here, I want to compare with Aristotle's account of the relationship between cognitive abilities and virtue to further understand Emerson's thoughts.

In Aristotle, human virtue is directly related to rationality, which he considers it as the primary function of human beings. For him, rationality takes two forms: theoretical rationality and practical rationality. The former is closely related to science, thinking about unchanging things, while the latter is thinking about changeable things. The good performance of the former function is intellectual virtue. Compared with Aristotle, Emerson does not emphasize this kind of rationality directly obtained through time cultivation and training. Emerson instead emphasizes intellectual intuition. The attainment of this intuition is closely related to a specific kind of devotion

---

[6] Ralph Waldo Emerson, *The Conduct of Life* (The Riverside Press, 1860), p. 152.

[7] Ralph Waldo Emerson, *Essays and Lectures*, ed. Joel Porte (Literary Classics of the United States, 1983), p. 417.

[8] Ibid.

[9] Ibid., p. 419.

which involves relinquishing the narrow cognition of self. The relevance of intellect and virtue is reflected at the level of will. Therefore, compared with Aristotle, Emerson's intellect has a religious connotation and it requires cognitive devotion to recognize the truth. This devotion is completed through faith. "By piety alone, by conversing with the cause of nature, is he safe and commands it. And because all knowledge is assimilation to the object of knowledge, as the power or genius of nature is ecstatic, so must its science or the description of it be."[10] In Emerson's view, with such devotion, people can better understand the truth. The truth as Emerson understands is not the objective scientific truth understood by Descartes, but the truth filled with value, which Emerson expresses with ecstasy. Therefore, the description of this truth must be lively and powerful, not dry facts.

Third, the cognitive faculties emphasized by Emerson also include imagination. Emerson mentions the concept of imagination in his essay titled "The Poet". "This insight, which expresses itself by what is called Imagination, is a very high sort of seeing, which does not come by study, but by the intellect being where and what it sees; by sharing the path or circuit of things through forms, and so making them translucid to others."[11] For Emerson, imagination is the unique ability of poets and it is superior to intellect. Poets can use the right words to transform the various material objects into the symbols of thought. For him, imagination has power in directing people to the truth. He writes, "But the quality of the imagination is to flow, and not to freeze. The poet did not stop at the color or the form, but read their meaning; neither may he rest in this meaning, but he makes the same objects exponents of his new thought."[12] For Emerson, imagination is the faculty of mind to perceive the truth of the world and it will not stop at the sensual qualities of the objects. Rather, it will try to discover the truth of the object.

Imagination also concerns the people and it can help people feel the reactions of others. Poets are the persons who possess this power. Imagination is closely related to other functions of mind. However, Emerson doesn't elaborate in depth about this relationship. Descartes also mentions imagination. For him, imagination is important part of the mind of human being. However, he defines imagination not in terms of its ability to perceive the unity rather in terms of its relationship with physical things in the world.

Regarding cognitive faculties, a very important point is that Emerson emphasizes the transformation of human beings' cognitive power. About transformation, he uses many similar concepts to explain, including tendency, saliency, transformation, progress, etc. Emerson emphasizes an important feature of the mind, which is the transformation of cognitive power. "The Universe is only in transit, or, we behold it shooting the gulf from past to future: And this the mind shares. Transition

---

[10] Ralph Waldo Emerson, *The Portable Emerson*, ed. Jeffrey S. Cramer (Penguin Books, 2014), p. 106.

[11] Ibid., p. 249.

[12] Ibid., p. 253.

is the attitude of power, and the essential act of life."[13] "In the instinct of progress, the mind is always passing—by successive leaps,—forward into new states, and, in that transition, is its health and power. The detachment which thought effects is the preparation for this step."[14] According to Emerson, the cognitive power of the human mind is also in the process of continuous progress. This advancing and transforming process is power. This power comes from the ultimate entity, that is, God. In addition, human cognition and natural changes correspond to each other. As nature continues to transform into new forms, the cognitive power of the human mind will have an innate response to these changes. So, why does the cognitive power of the human mind have an innate response to natural changes? This is closely related to the value of nature. The objects of nature have the value of truth, goodness, and beauty. These forces naturally inspire the human mind to understand nature. He writes, "The moral influence of nature upon every individual is that amount of truth which it illustrates to him."[15] For Emerson, there is truth in nature, which is also termed as moral law, or spiritual law. It has the power to influence people's mind. We cannot choose to resist its influence.

1.2 Virtue in Confucianism: Emotional Rationality, Power, and Values

From the above, it is evident that Emerson's concepts of cognitive abilities including reason, understanding, intellect, and imagination are primarily directed towards comprehending the natural world and the universe. For Emerson, these faculties orbit around the pursuit of universal truth of existence. In contrast, Confucianism also stresses reason, which is deeply related to understanding and imagination. However, in Confucian classical texts, there is no such concept of reason. Instead, the concept of *Heart and Mind* (xin) is frequently used. Professor Peiyuan Meng's characterization of Heart and Mind is very insightful. For him, the Confucianism's concept of Heart and Mind is in essence "emotion-oriented reason."[16] He puts forward the concept of emotion-oriented reason in this passage: "The Confucian concept of *cheng* (sincerity) is not only an expression of moral emotion but also a manifestation of heavenly principles. Here, emotion and reason converge into a distinctive 'emotion-oriented reason'."[17] The concept of *cheng* in this passage is often combined with the words of *Heart and Mind* in Chinese language. People use it very often in everyday life. Cheng as the key moral term is used to represent the main feature of *Heart and*

[13] Ralph Waldo Emerson, *The Best Read Naturalist*, ed. Michael P. Branche and Clinton Mohs (University of Virgina Press, 2017), p. 216.

[14] Ibid., p. 180.

[15] Ralph Waldo Emerson, *The Portable Emerson*, ed. Jeffrey S. Cramer (Penguin Books, 2014), p. 36.

[16] Meng Peiyuan, *Emotion and Reason* (Beijing: China Renmin University Press, 2009), pp. 10–30.

[17] Ibid.

Mind. What, then, is emotion-oriented reason? Professor Meng Peiyuan doesn't give us a philosophical analysis in depth. In this part, I will explore this concept deeply through three philosophical dimensions: (1) understanding;(2) self-control;(3) value and power.

First, emotional-oriented rationality entails the capacity to comprehend human emotions, which are profoundly shaped by the practice of rites. Unlike Emerson's emphasis on understanding the physical world and its cosmic relationships, Confucian understanding is oriented toward the world of human beings. Within this context, it focuses on the knowledge of individuals including their intimate relationships, social roles, psychological states, feelings, and all factors influencing their well-being in their lifetime. Confucianism particularly prioritizes the innate affection people feel toward their parents, viewing these emotions as possessing a moral dimension. This reverence is encapsulated in the concept of filial piety (xiao 孝), which Confucianism regards as a primary moral obligation.

Second, emotional rationality involves self-control. It restrains the self from deviating from traditional rites or pursuing goals unrelated to fundamental human duties. In Confucian thought, the primary responsibilities for human beings are to fulfill obligations toward parents, siblings, and friends. To neglect these duties in favor of pursuits such as seeking universal truth as Emerson advocates would be deemed morally wrong. Moreover, self-control in this framework is not merely an abstract virtue but is cultivated through the consistent practice of rites. Without this practical engagement, understanding moral principles alone is insufficient to achieve self-discipline.

Third, emotional rationality is guided by value. The core values are benevolence and righteousness. The primary function of rites in Confucianism is to foster and sustain this harmony. These values are closely related to power. This dynamic relationship between values and power is vividly illustrated in a well-known passage from Mencius: "As power, it is exceedingly great and exceedingly strong. If nourished by uprightness and not injured, it will fill up all between heaven and earth. As power, it is accompanied by righteousness and the Way. Without them, it will be devoid of nourishment.'"[18] This passage reveals that, for Mencius, emotional rationality is intimately tied to the will. When the will is focused, it cultivates a power that aligns with Heaven and Earth. This concentration, however, depends on the guiding values of righteousness; without such values, achieving this unity becomes difficult.

Through the above explanation of the ability of Confucian emotional rationality, we can see several main differences between Confucianism and Emerson in the understanding of cognitive power. (1) Confucian emotional rationality regards interpersonal emotions as morally good and pursues it in the form of rites. Emerson emphasizes the moral sentiment between man and God, which is the source for moral sentiment among people. (2) Confucianism's argument for moral emotions focuses on experientiality, that is, starting from observations in life to find what

---

[18] *A Source Book in Chinese Philosophy*, trans. Wing-Tsit Chan (Princeton University Press, 1963), p. 63.

people commonly pursue. Emerson, on the other hand, focuses on religiosity, emphasizing the direct inspiration of moral emotions to the soul. (3) Confucianism focuses on cultivating habits through rites to achieve the core virtue of benevolence while Emerson opposes the constraints of secular norms and emphasizes the changes in actions brought about by a direct understanding of God.

1.3 Virtue in Daoism: Reflective Reason, Power, and Values

In Daoism, I will mainly focus on Zhuangzi. As I mentioned earlier, the core virtue for Zhuangzi is wandering. The virtue of wandering is the result of reflective reason. There is one thing I would like to stress here, which is that Zhuangzi never uses the concept of reflective reason; rather, he uses the concept of an empty mind. An empty mind is the literal translation for the Chinese words *xu xin* (虚心). However, this translation tends to be misleading as it cannot capture Zhuangzi's philosophical ideas on the mind. An empty mind can be best explained as reflective reason. I will address reflective reason from the following three aspects: (1) understanding; (2) choice and control; (3) values and powers.

First, reflective reason is based on understanding. In terms of understanding, Zhuangzi states, "'To understand what is done by Heaven, and also what is to be done by Man, that is the utmost.' To understand what is done by the Heavenly: just in being the Heavenly, as the way all beings are born, what it does is give birth to them all. To understand what is to be done by the Human: that would be to use what your understanding understands to nurture what your understanding does not understand. You could then live out all your natural years without being cut down halfway. And that would indeed be the richest sort of knowledge."[19] Zhuangzi and Emerson both use the concept of understanding in their works. But their interpretation of understanding is different. I would like to examine the difference from the following three aspects: (1) For Emerson, understanding is directed to the truth of the material world. For Zhuangzi, understanding is not directed to the truth of the material world, but it is directed to the world of values. The core value Zhuangzi advocates is spontaneity, or being self-so, which is just the opposite of man-made value. Man-made values are characterized by purposeful pursuit. For example, the purposeful pursuit of Confucian moral values is one type of man-made value. (2) Zhuangzi believes that humans cannot obtain the truth of objective things because everything is constantly changing. This is exactly the opposite of Emerson. Emerson believes that nature is always moving and evolving in a better direction, and human cognition will also improve accordingly, and humans will eventually obtain the truth of all things; (3) Zhuangzi's concept of understanding is about self-knowledge. This knowledge includes recognizing the prevailing values of the time, mainly manifested as Confucian benevolence and righteousness, which have a huge impact on

[19] Zhuangzi, *Zhuangzi: The Essential Writings with Selections from Traditional Commentaries*, trans. Brook Ziporyn (Hackett Publishing Company, 2009), p. 39.

human emotions and actions. Zhuangzi saw the negative impact of many individuals' emotions and actions being shaped by the values of benevolence and righteousness in society at that time, including the importance attached to life and the disgust of death. It also includes the harm to life brought about by the pursuit of benevolence, righteousness, and fame, and the hypocrisy of emotions brought about by overemphasis on compliance with rituals. Having sufficient knowledge is a prerequisite for reflective rationality.

Second, reflective rationality includes choice and control. By choice, I mean actively choosing a value under the condition of judging various values. What Zhuangzi emphasizes as non-action seems to conflict with choice because choice implies the power of rational thinking in action, while non-action seems to represent the abandonment of action. But Zhuangzi's non-action is a natural state that arises from choosing not to choose. It includes a process of choice. Free and easy wandering is the result of choosing not to choose. It is not the original state before any choice. Reflective rationality is also related to control. By control, I mean we control our inner mind and achieve inner harmony without being disturbed by external changes in various events.

Third, reflective rationality includes the dimension of value. When people choose not to choose, it is still guided by the value of spontaneity or being self-so. Zhuangzi regards being self-so as good, something that can bring vitality to people. Reflective rationality always revolves around this core value. Zhuangzi's reflective rationality refers to the reflection on various moral values. This reflection is based on having sufficient knowledge and life experience. Zhuangzi mentioned a lot of Confucian thoughts and the various practices brought about by Confucianism including political practices, etc. Zhuangzi reflects on these thoughts and practices on the basis of a full understanding. After reflection, he does not set out to establish a new moral value. Instead, he transcends the existing moral values. Compared with Emerson, he emphasizes piety more at the cognitive level. Emerson does not reflect on the ultimate value, that is, the truth, goodness, and beauty of God, but confirms its existence through faith.

2 Virtue: Bodily Power and Values

The concept of virtue in Emerson and Chinese philosophy is closely related to bodily power and value. In terms of bodily power and its relationship with value, the basic logic can be expressed as follows: The body has the moral power that will influence other people's values and thoughts. Values can influence bodily power, which is demonstrated by bodily expressions. In the following, I will examine bodily power and its relationship with value from three dimensions: (1) bodily expressions when interacting with other people; (2) bodily interaction with the land; (3) the philosophical status of the body.

2.1 Virtue in Emerson: Bodily Power and Values

Emerson's concept of body is deeply related to virtue, power and values. In this section, I will discuss his concept of body from the following three aspects: The first is the state presented when people interact with each other, which includes language, expressions, gaze, and actions. The second aspect is the state that the body presents when interacting with nature, including observing nature, working on the land, and various sports activities. The third aspect is about the philosophical status of the body. At the same time, I will discuss the interrelationship among virtue, bodily power, and values.

First, Emerson's concept of body has the meaning of the states people exhibit in their interactions with one another, which is termed as manners. In his essay "Behavior", Emerson gives a defintion of manners. Emerson writes, "The visible carriage or action of the individual, as resulting from his organization and his will combined, we call manners."[20] The concept of will in this passage doesn't refer to self-will which he thinks represents the false self. The will refers to the will that have been surrendered to God. By visible carriage or action, he refers to form, attitude, gesture, facial expressions, and so on. For him, values in our mind will have direct influence on visible bodily state. He also stresses the power of manners. The term "power" is used many times by him in his essay when he explains manners. "Manners impress as they indicate real power. A man who is sure of his point, carries a broad and contented expression, which everybody reads."[21] Manner for Emerson is the external bodily expression which is the result of human nature, which is in essence the embodiment of ultimate values. It has power to influence other people's values and opinions.

Second, the interaction between the body and the land manifests in two distinct forms: farming and outdoor sports. Regarding the former, Emerson underscores the value of manual labor as a foundation for higher pursuits. He points out that a man should have a farm or a mechanical craft for his culture. We must have a basis for our higher accomplishments, our delicate entertainments of poetry and philosophy, in the work of our hands.[22] For the latter, Emerson explores the merits of physical activities like sports, which he deems more beneficial to him than the monotonous labor of farmers and craftsmen.[23] Yet, he views all bodily powers as subordinate to the mind, serving as essential mediums for spiritual growth. While he acknowledges the reliability of those engaged in physical toil, he notes their typical lack of organizational capacity. Emerson places greater emphasis on the power of the mind. This power encompasses reason, imagination, and intellectual cultivation. He likes to call it "a new gymnasium." He writes, "We go to the gymnasium and the swimming-school to see the power and beauty of the body; there is the like pleasure, and a higher benefit, from witnessing intellectual feats of all kinds; as, feats of memory, of

---

20 Ralph Waldo Emerson, *The Conduct of Life* (The Riverside Press, 1860), p. 147.

21 Ibid., p. 164.

22 Ralph Waldo Emerson, *Essays and English Traits* (New York: P. F. Collier, 1909), p. 50.

23 Ibid. p. 59.

mathematical combination, great power of abstraction, the transmutings of the imagination, even versatility, and concentration, as these acts expose the invisible organs and members of the mind, which respond, member for member, to the parts of the body. For, we thus enter a new gymnasium, and learn to choose men by their truest marks, taught, with Plato, 'to choose those who can, without aid from the eyes, or any other sense, proceed to truth and to being.'"[24] From these passages, we can see that for Emerson, our bodily engagement with the land is not valuable for its own sake. For him, the metaphysical status of body is inferior to the mind. Although he doesn't ignore the body philosophically as Plato has done in his dialogues, he still treats body as the medium for people to develop the mind. He writes, "Without looking, then, to those extraordinal social influences which are not acting in precisely this direction, but only at what is inevitable doing around us, I think we must regard the *land* as a commanding and increasing power on the citizen, the sanative and Americanising influence, which promises to disclose new virtues for generations."[25] For Emerson, developing the relationship with land can cultivate human beings' new virtues.

Third, regarding the philosophical status of the body, Emerson acknowledges its role in experiencing the world but assigns it a secondary position. For him, the direct bodily experience of nature is the first step for people to know the transcendental truth in God. However, in Emerson's philosophy, the body doesn't have an important philosophical status as the spirit, which is the creator of the body and all things. Emerson, to some extent, also inherits the legacy of Western philosophical tradition. He believes that there is a hierarchy between the body and the spirit, with the spirit being superior to the body. The ultimate return to nature is not to coexist with nature in the body but to return to a kind of spirit. However, in Emerson's view, although the body is metaphysically inferior to the spirit, Emerson doesn't regard the body as a negative force that hinders the soul from approaching the truth like many Western philosophers. Instead, he emphasizes the mediating role of the body. The body, in his view, is merely a conduit for developing the mind and apprehending God. He observes, "To be sure, there are men enough who are immersed in things, as farmers, smiths, sailors, and mechanics generally; and we know how real and solid such men appear in the presence of scholars and grammarians: but these men ordinarily lack the power of arrangement, and are like hands without a head."[26] Here, bodily powers do not define human essence but are necessary for mental development. In this respect, Emerson aligns with Western philosophers like Aristotle and Kant, for whom physical capacities are valuable only insofar as they enhance intellectual faculties.

---

[24] Ralph Waldo Emerson, *Essays and Lectures*, ed. Joel Porte (Literary Classics of the Untied States, 1983), p. 622.

[25] Ralph Waldo Emerson, *English Traits & Representative Men & Other Essays*, ed. Ernest Rhys, London: J. M. Dent & Sons, Ltd, 1908), p. 362.

[26] Ibid., p. 730.

2.2 *Virtue in Confucianism: Bodily Power and Values*

The Chinese word for the body is the combination of two Chinese characters, which are "身" and "体". The word "body" (*shen*) appears about 18 times in *The Analects* 54 times in *Mencius*. This concept is more closely related to power and value. In the following, I will examine the relationship among bodily power, virtue, and values in Confucianism from three aspects: (1) I will examine the bodily expression when we interact with each other and virtue; (2) Our bodily engagement of the land and its relationship with virtue; (3) the philosophical status of body.

First, in terms of the bodily expression when we interact with each other, Confucianism stresses propriety, which governs almost all the aspects of everyday life including our facial expressions, emotions, dressing, eating, and so on. For Confucianism, it is the external expressions of the core virtue of benevolence. There are many passages in Confucianism on this aspect. In terms of external bodily expressions, the big difference between Emerson and Confucianism is that the former stresses God-inspired feelings of wonder, while Confucianism stresses the harmony in feelings. Harmony refers to the equilibrium of mind. In *the Doctrine of the Mean*, a passage states, "Before the feelings of pleasure, anger, sorrow, and joy are aroused it is called equilibrium (*Chung*, centrality, mean). When these feelings are aroused and each and all attain due measure and degree, it is called harmony."[27] For Confucianism, all of these feelings are related to facial expression and demeanor. If we control the extreme feelings and are moderate in the expression of these feelings, our bodily power will influence people.

Second, in terms of our bodily engagement with the land, Confucianism stresses the foundation of working on land for developing virtue. For example, Mencius stresses that agricultural activities on the land can help us develop our virtue. "The way according to which the people conduct their lives is this: If they have a secure livelihood, they will have a secure mind. If they have no secure livelihood, they will not have a secure mind. And if they have no secure mind, there is nothing they will not do in the way of self-abandonment, moral deflection, depravity, and wild license."[28] In Mencius' lifetime, agriculture was the main livelihood. Cultivating the virtue needs to have secure living through working on the farming land. Bodily engagement with the land for Confucianism is the foundation for us to develop the virtue. In this aspect, Mencius is similar to Emerson in that they both stress bodily interactions of the land and its instrumental role in developing virtues.

Thirdly, Body occupies an important philosophical status in Confucianism. When Confucianism mentions the body, it often combines it with characters such as "examine", "peace", and "cultivate". For example, in *The Analects,* there are expressions about the body. Tseng Tzu said: "Every day I examine myself on three counts. In what I have undertaken on another's behalf, have I failed to do my best? In my dealings with my friends have I failed to be trustworthy in what I say? Have I passed on to

[27] *A Source Book in Chinese Philosophy*, trans. Wing-Tsit Chan (Princeton University Press, 1963), p. 98.

[28] Ibid., p. 66.

others anything that I have not tried out myself?"[29] The body in Confucianism usually doesn't refer to the body in the biological sense; rather, it is related to the moral self. In Confucianism, the body and virtue are inseparable. Whether it is Confucianism or Daoism, their understanding of the body is consistent in this respect: they all believe that the basic constituent of the body is *qi*, which is the same as the composition of other natural objects outside the body. The main characteristic of *qi* is fluidity. When *qi* is in the best state of flow in the body, people generate special cognitive abilities that allow them to recognize the ultimate ontology. The flow of *qi* is influenced by values, which are manifested as benevolence and righteousness in Confucianism and as nature in Daoism. Confucianism holds that when people have the values of benevolence and righteousness, the flow of *qi* in the body will be in the best state, and the integration with the *qi* of Heaven and Earth will be realized. Daoism, on the other hand, believes that when people get rid of the values of benevolence and righteousness and are in a state of natural tranquility, they can achieve integration with heaven and earth. For example, Mencius writes in the chapter Gongsun Chou.

> Ch'ou asked, 'May I venture to ask, sir, in what you are strong?'
>
> Mencius replied, 'I understand words. And I am skillful in nourishing my strong, moving power.'
>
> 'May I ask what is meant by the strong, moving power?'
>
> 'It is difficult to describe. As power, it is exceedingly great and exceedingly strong. If nourished by uprightness and not injured, it will fill up all between heaven and earth. As power, it is accompanied by righteousness and the Way. Without them, it will be devoid of nourishment. It is produced by the accumulation of righteous deeds but is not obtained by incidental acts of righteousness. When one's conduct is not satisfactory to his own mind, then one will be devoid of nourishment. I therefore said that Kao Tzu never understood righteousness because he made it something external.'[30]

From this passage, we can see that the power of the body as understood by Mencius is the power of *qi* itself, which can allow the *qi* of the body to merge with the *qi* of heaven and earth. However, to achieve this integration, it requires the power brought by *qi* itself, which can only be maximized under the drive of values, and these values are benevolence and righteousness.

2.3 Virtue in Daoism: Bodily Powers and Values

In Daoism, the basic logical relationship among bodily power, value, and virtue can be expressed as follows: Value can strengthen bodily power. The value that Daoism stresses is spontaneity. When bodily power is strengthened, we will have virtue. In the following, I will address this relationship from the three aspects: the bodily expression we exhibit when we interact with others and virtue; bodily interaction

[29] Confucius, *The Analects*, trans. D. C. Lau (Penguin Classics, 1979), p. 59.

[30] *A Source Book in Chinese Philosophy*, trans. Wing-Tsit Chan (Princeton University Press, 1963), p. 63.

with the land and virtue; the philosophical status of body and its relationship with value and virtue.

First, in terms of bodily expression we exhibit when we interact with others, the chapter "The Sign of Virtue Complete" in Zhuangzi best expresses his view of virtue through the mouth of Confucius. The Chinese expression for "whole in power" here is 才全 *cai quan*, which is closely related to virtue. When Zhuangzi explains the words, he uses "virtue takes no form". It is just opposite to Confucianism. For Confucianism, virtue will take form which is reflected in rituals. The rituals will have a lot of requirements on the bodily expression. Zhuangzi expresses his strong criticism here. "Likewise, in the days of Hexu, the people stayed at home without knowing what they were doing, ventured out without knowing where they were going. Filling their mouths they were merry; drumming their bellies they amused themselves. This was the extent of the people's abilities. Then along came the sage, bending and twisting over ritual and music to reform the bodies of the world, dangling Humanity and Responsibility overhead "to comfort the hearts of everyone in the world."[31] For Confucianism, virtues will require rites, which are closely related to reforming the bodies including dressing, the appearances, demeanor, eating, and paying respect towards other people and so on. However, for Zhuangzi, if we have virtue, the main feature of external appearance is tranquility without being disturbed by changes in events.

Second, in terms of our interaction with the land, Zhuangzi stresses interaction with mountains and forests and also working on the agricultural land. For him, the two ways both represent a person's virtue. The following passages can best explain Zhuangzi's view of bodily interaction with land or forest and its relationship with virtue. "When there is insufficient space in a house, the womenfolk will surely fall to quarreling. If the mind has no Heavenly wandering, the six apertures of perception will interfere with one another. This is why being in a vast mountain forest benefits a man more than whatever might be granted by the gods."[32] Mountain and forest are often jointed to use together to represent nature with vast spaces, which will contribute to development of our virtues. Another way of interaction with the land is to work on the land as a farmer. For Zhuangzi, working in a agricultural land represents human nature, which is deeply connected to virtue. "For the people too have their own constant inborn nature. To be clothed by their own weaving, fed by their own plowing —this is what is called their shared Virtuosity."[33] For Zhuangzi, working on the farmland and getting our basic living materials is inborn human nature and virtue is deeply related to this mundane life.

Third, in terms of the philosophical status of the body, the following passages from Laozi and Zhuangzi can provide us a deep insight.

> Which does one love more, fame or one's own life?
>
> Which is more valuable, one's own life or wealth?

---

[31] Zhuangzi, *Zhuangzi: The Essential Writings with Selections from Traditional Commentaries*, trans. Brook Ziporyn (Hackett Publishing Company, 2009), p. 62.

[32] Ibid., p. 113.

[33] Ibid, p. 61.

Which is worse, gain or loss?

Therefore he who has lavish desires will spend extravagantly.

He who hoards most will lose heavily.

He who is contented suffers no disgrace.

He who knows when to stop is free from danger.

Therefore he can long endure.[34]

Let me try explaining what I mean. From the Three Dynasties on down, everyone in the world has altered his inborn nature because of some [external] thing. The petty man?—he will risk death for the sake of profit. The knight?—he will risk it for the sake of fame. The high official? –he will risk it for family; the sage? –he will risk it for the world.[35]

Those in ancient times who wished to keep themselves alive did not use eloquence to ornament their knowledge. They did not used their knowledge to make trouble for the world; they did not use their knowledge to make trouble for virtue. Loftily they kept to their places and returned to their inborn nature.[36]

In the above three passages, the English expressions including "one's own life", "risk death", and "themselves" are all translations for the Chinese philosophical concept of body. For Daoism, body can represent the essence and identity of human beings. However, the pursuit of fame, virtue, or wealth will hurt our human essence.

I would like to explain the status of body in Zhuangzi from three aspects: (1) divine and spiritual; (2) the object of cognition and bodily power; (3) cultivation of the body. There are several points to emphasize about the divine and the spiritual: (1) The divine and the spiritual mentioned here are not a transcendental ontology, not God as Emerson understands, but cognition closely related to the human body. It is closely related to the flow of *qi* in the human body. When the mind is in a state of emptiness, that is, when the false self is cleared from the mind, then the flow of *qi* in the human body will be in the best state, and people can better realize a kind of intuition. This kind of cognition is usually inexpressible in language; (2) The object of cognition is not the material world. In Zhuangzi's view, the external world is constantly changing, and we cannot obtain objective knowledge of the external world. The object of cognition he emphasizes mainly refers to the cognition of the object involved in human skills; (3) The acquisition of this cognitive power requires the power of the individual's will within the mind, which requires a long process. Unlike Emerson, this power of will is not guided by the highest ontology, the Dao, because the Dao itself has no purpose. Therefore, compared with Emerson, Zhuangzi emphasizes that the acquisition of cognitive power requires internal struggle of the mind, but Emerson does not emphasize this struggle. In his view, a person can become a virtuous person in an instant. People with this cognitive ability are often referred to by Zhuangzi as *zhen ren* (authentic person), which corresponds to true knowledge, but this true knowledge is not the scientific truth described by Emerson, but the true

[34] *A Source Book in Chinese Philosophy*, trans. Wing-Tsit Chan (Princeton University Press, 1963), p. 161.

[35] Zhuangzi, *The Complete Works of Chuang Tzu*, trans. Burton Watson (Columbia University Press, 1968), p. 101.

[36] Ibid., p. 174.

knowledge of the Dao in the field sense. When Zhuangzi describes this cognitive ability, he often uses the sill abilities of some crafsmen as examples to explain. But this is a specific sill cognition. In Aristotle's view, this cognition represents skill, and sill is one of the categories of human rational virtues, although this category is in a lower position compared to human scientific cognition. It can be seen that Emerson emphasizes more on people who can perceive the essence of natural spirit at the cognitive level, rather than cognitive abilities developed in specific skills.

3 Virtue: The Power of Action and Values

The power of action I am referring to is the power of moral action, that is, the power to engage in various moral practices. Both Emerson and Confucianism emphasize the power of action, while Zhuangzi focuses more on the spiritual level of free and easy wandering, which is an aesthetic attitude, which does not motivate the moral action. Confucianism focuses more on direct moral action rather on rational reflection on the rational for the moral action. Confucianism focuses on specific practices, rather than reflecting on the rationality of the practice itself. Zhuangzi focuses on aesthetic attitude which is also regarded as moral virtue termed as "wandering" (逍遥游). He does not regard Confucian moral virtues as the only criteria for self-identity. He advocates transcending the criteria. In terms of the philosophical status of the power of action, Emerson is between Confucianism and Daoism. Philosophical reflection of action is more superior to direct moral action. For Emerson, knowing the truth is most important and it can better define ourselves. Therefore, compared with Confucianism, Emerson has more openness in understanding values. He will not regard a value as the ultimate, but views value with a constantly changing and developing perspective. But Confucianism doesn't question tradition. Rather, it will inherit tradition. From this point of view, Daoism and Emerson have similarities. Zhuangzi emphasizes that people do not have a fixed identity, what he mainly refers to is not to be constrained by a fixed moral value, emphasizing the changeability of self-identity. But the biggest difference between Zhuangzi and Emerson is that Emerson regards the ultimate ontology, nature, as having a certain tendency, which is to constantly develop in a better direction. However Zhuangzi emphasizes is natural, this inevitable force to transform in a better direction does not exist. This also explains why Emerson emphasizes action. But Emerson will not immerse himself in action, but constantly think, to get closer to the truth.

3.1 Virtue in Emerson: Power of Action and Values

Emerson's concept of virtue is also related to power of action and values. Although the power of action is less important compared with the power of the mind. Emerson expressed his views on action and power in the his essay "Experience" and "The

Poet". Emerson writes, "…the true romance which the world exists to realize will be the transformation of genius into practical power."[37] Also, in the essay titled the American Scholar, he writes, "The true scholar grudges every opportunity of action past by, as a loss of power."[38] It can be seen that Emerson regards action as the raw material for creating thoughts, and action itself is not the ultimate goal, it is just a medium to generate thoughts. The power of action includes the power to improve social injustice, such as his own public opposition to slavery, which is a type of power of action. In terms of the power of action, a question arises here: what is the source of it? As mentioned earlier, Emerson thinks that virtue is constantly changing, and there is no ultimate virtue. All are initial and will be surpassed by later virtues. So, does this cognition hinder people's power of action? For this question, a category used by Emerson can provide a good answer. That is moral sentiment. He has defined moral sentiment in his essay of compensation, worship and in his speech on morals. "Thus is the universe alive. All things are moral. That soul, which within us is a sentiment, outside of us is a law. We feel its inspiration; out there in history we can see its fatal strength. 'It is in the world, and the world was made by it.'";[39] "The law is the basis of the human mind. In us, it is inspiration; out there in Nature, we see its fatal strength. We call it the moral sentiment."[40] Emerson's explanation of moral sentiment, there are several aspects I want to further explain. (1) Emerson's moral sentiment is top-down, it is the flow of the highest ontology, God (also known as the soul), to the human mind. Humans can feel the inspiration brought by this sentiment. (2) The moral sentiment represents the core of human nature. It can help us establish a good relationship with nature and connect us with the ultimate cause. This connection is inseparable from science. This moral sentiment comes from the belief in the ultimate ontology, that is, God. Moral sentiment will constantly inspire people to discover the truth, and this truth also controls the moral world of mankind. The power of action comes from the intuitive cognition of truth, this cognition will be constantly in a state of promotion, until it finally approaches the truth.(3) Moral sentiment is deeply connected to beauty. Emerson explains the beauty in connection with sublime feelings. When moral sentiment reveals truth to the human mind, we will be full of aesthetic excitement.

In addition, Emerson opposes people pursuing action itself as the ultimate goal. He expresses this view in his essay "Circles". He writes, "Our moral we have already deduced in considering the circular or compensatory character of every human action. Another analogy we shall now trace, that every action admits of being outdone."[41] Emerson opposes people treating action itself as the ultimate goal, complacently

---

[37] Ralph Waldo Emerson, *The Portable Emerson*, ed. Jeffrey S. Cramer (Penguin Books, 2014), p. 278.

[38] Ibid., p. 63.

[39] Ralph Waldo Emerson, *Essays and Lectures*, ed. Joel Porte (Literary Classics of the United States, 1983), p. 289.

[40] Ibid., p. 891.

[41] Ralph Waldo Emerson, *The Portable Emerson*, ed. Jeffrey S. Cramer (Penguin Books, 2014), p. 226.

stopping progress within the action, and instead emphasizes the truth itself. A single action cannot prove truthfulness, because the truth understood by Emerson is a truth that is constantly advancing. When people act, it is not the action made by rational reasoning and judgment as understood by Kant, but rather seeing the individual as the ultimate entity, that is, the channel through which God conveys his truth. God is always communicating with the human heart in various languages. Humans only need to accept themselves as a channel for transmitting the voice of God. His actions are also unfolded in the intuition of the revelations that God brings to the human soul. Therefore although it is an uncertain perception, it will carry the power of beauty which can motivate people to engage in moral action. Emerson expressed this well in his essay "The Transcendentalist", "But this class are not sufficiently characterized if we omit to add that they are lovers and worshippers of Beauty. In the eternal trinity of Truth, Goodness, and Beauty, each in its perfection including the three, they prefer to make Beauty the sign and head."[42] Emerson emphasizes beauty because beauty can well avoid the obsolescence of goodness and the ruthlessness of truth, and beauty can promote better action.

3.2 *Virtue in Confucianism: Power of Action and Values*

In contrast to Ralph Waldo Emerson who focuses more on cognitive power rather than power of action, Confucianism focuses more on action. The key concept for us to better understand Confucianism' action oriented virtue is "thinking" (si思). Thinking is the translation for the Chinese word 思. This translation is very misleading since Confucian "thinking" encompasses moral activities—emotions, intentions, and behaviors—rather than mere intellectual cognition. To clarify this, we'll examine key passages from Confucian texts, beginning with Mencius, followed by an analysis of "thinking" in *The Analects*, to illuminate its role in shaping moral action. Mencius explains the relationship between the heart and thinking as follows: "'He who is guided by the interests of the parts of his person that are of greater importance is a great man; he who is guided by the interests of the parts of his person that are of smaller importance is a small man.' 'Though equally human, why are some men guided one way and others guided another way?' 'The organs of hearing and sight are unable to think and can be misled by external things. When one thing acts on another, all it does is to attract it. The organ of the heart can think. But it will find the answer only if it does think; otherwise, it will not find the answer. This is what Heaven has given me. If one makes one's stand on what is of greater importance in the first instance, what is of smaller importance cannot usurp its place. In this way, one cannot but be a great man."[43] Mencius distinguishes between the sensory faculties

[42] Ralph Waldo Emerson, *The Portable Emerson*, ed. Jeffrey S. Cramer (Penguin Books, 1946), p. 125.

[43] Mencius, *Mencius*, trans. D. C. Lau (Penguin Classics, 2004), p. 65.
Confucius, *The Analects*, trans. D. C. Lau (Penguin Books, 1970), p. 168.

(ears and eyes), which passively react to external stimuli without discernment, and the heart, the seat of moral reflection. "Thinking" here is not abstract reasoning but an active, moral engagement that discerns right from wrong. The "great parts" refer to the heart's capacity for virtue, which, when cultivated, elevates one to a "great man", while neglect leads to pettiness. This underscores that Confucian "thinking" is a function of the heart, not the mind in a Western philosophical sense.

The Confucian understanding of "thinking" is further refined in *The Analects*: "The master said, 'If one learns from others but does not think, one will be bewildered. If, on the other hand, one thinks but does not learn from others, one will be in peril'."[44] In *The Analects*, "thinking" complements learning and action, forming a balanced moral practice. The quote warns against unreflective study or ungrounded reflection, emphasizing that "thinking" integrates knowledge into ethical behavior. Together, these passages frame "thinking" as a practical, heart-centered process. Mencius states, "Even with a *t'ung* or a *tzu* tree one or two spans thick, anyone wishing to keep it alive will know how it should be tended, yet when it comes to one's own person, one does not know how to tend it. Surely one does not love one's person any less than the *t'ung* or the *tzu*? This is unthinking to the highest degree."[45] This passage illustrates "thinking" as a deliberate act of moral care. Mencius criticizes the failure to nurture one's moral self as a lack of "thinking", contrasting it with the care given to trees. Here, "thinking" is an intentional effort to align actions with virtue, not a passive intellectual exercise. Unlike Emerson, who links action to an intuitive pursuit of divine truth, Confucianism views action as independent of such metaphysical aims, rooting it instead in the concept of "thinking". The English translation "think" misleads, as it suggests Western rational cognition, whereas Confucian "thinking" denotes moral activities of the heart—encompassing emotions, intentions, and behaviors. To understand this, let's explore key Confucian texts.

In Confucianism, "thinking" is not about uncovering universal truths, as in Emerson's transcendentalism, but about sustaining moral order through the heart's activity. It drives action not toward an ultimate truth but toward harmony within human relationships and society. This cognitive capacity, rooted in moral emotion rather than intellectual abstraction, defines Confucian action as a lived, ethical practice.

3.3 *Virtue in Daoism: The Power of Action and Values*

Zhuangzi's "sitting in oblivion" is the core category for understanding the inner action of the mind. Many sections related to skills in Zhuangzi's works, when describing the reasons for a person's mastery of skills, they all attribute the source to oblivion, and this oblivion often takes several days to complete. The "several days" here does not refer to the specific number of physical days, but refers to the fact that this

---

[44] Confucius, *The Analects*, trans. D. C. Lau (Penguin Classics, 2004), p. 65.

[45] Mencius, *Mencius*, trans. D. C. Lau (Penguin Books, 1970), p. 167.

process requires long-term persistence to complete. For example, In the chapter "The Great and venerable Teacher" chapter, the dialogue between Yan Hui and Confucius, Zhuangzi mentioned Yan Hui's two processes of forgetting, namely forgetting rites and music and forgetting benevolence and righteousness, but still did not reach the "sitting in oblivion" that Zhuangzi wanted to express through Confucius' mouth. Zhuangzi expresses the idea of "sitting in oblivion" as follows: "It's a dropping away of my limbs and torso, a chasing off of my sensory acuity, which disperses my physical form and ousts my understanding until I am the same as the Transforming Openness."[46] Forgetting things that are irrelevant to the essence of life, in Zhuangzi's view, is not completed through any religious rituals, nor through the cultivation of the rational self, but through the inner effort of the mind. But in Emerson's view, his inner action does not require repeated efforts like Zhuangzi's, He does not emphasize the struggle of the heart, but emphasizes the power of faith, which will reduce the internal struggle of people. He writes, "Place yourself in the middle of stream of power and wisdom which animates all whom it floats, and you are without effort impelled to truth, to right and a perfect contentment. Then you put all gainsayers in the wrong. Then you are the world, the measure of right, of truth, of beauty."[47]

In Emerson's view, when we place ourselves in the torrent of truth, goodness, and beauty through the power of faith, we naturally possess truth, goodness, and beauty. But in Zhuangzi's view, he does not believe that there is such an ultimate truth, goodness, and beauty, because his concept of Dao is beyond the standard of good and evil. And the operation of Dao is natural. Zhuangzi is passive in external actions compared to Emerson, he emphasizes not cooperating in politics. He will not strive to make the world better through existing political power. He emphasizes the virtue of wandering and a individual aesthetic life style.

Through the comparison of Emerson and Zhuangzi's views on action, we can see a significant difference, that is, Emerson emphasizes the action unfolded in society, facing unjust events, he will stop it through public speeches or writing. But Zhuangzi tends to not comment on social injustices, and will not hinder social decay through political efforts, that is, persuading the king. One of the main reasons is the difference in values. Although Emerson believes that human morality is constantly changing, and new values are constantly replacing old values, he does not deny the truth, goodness, and beauty of current values. This value is his main motivation for action. In the chapter on transcendentalists, he once pointed out that their call for justice for blacks, beggars, and drunks is for beauty, a spiritual need of the caller. They did not use morality as a driving force, because of the instability of morality, "We call the Beautiful the highest, because it appears to us the golden mean, escaping the dowdiness of the good and the heartlessness of the true. They are lovers of nature also, and find an indemnity in the inviolable order of the world for the violated order and grace of man."[48] From this passage, we can see that Emerson stresses the motivation

---

[46] Zhuangzi, *Zhuangzi: The Essential Writings with Selections From Traditional Commentaries*, trans. Brook Ziporyn (Hackett Publishing Company, 2009), p. 49.

[47] Ralph Waldo Emerson, *The Portable Emerson* (Jeffrey S. Cramer, Penguin Books, 2014), p. 195.

[48] Ibid. p. 125.

of moral actions from beauty rather than the sense of responsibility as emphasized by Confucianism. In Confucianism, these responsibilities are mostly oriented towards close relationships. Moral consideration and responsibilities towards people outside these relationships tend to be weak. However, Emerson's stress on the motivation from beauty can generate new moral responsibilities towards people who are outside close relationships, especially the underrepresented groups in society. This beauty will drive them to discover new moral values, that is, moral goodness. Beauty has a guiding role, it will inspire a person's intuition to do the right thing. Compared with Emerson, in Zhuangzi, there is no objective spiritual beauty to guide people to take action. For various external social problems, Zhuangzi tends to pursue inner peace and will not value the various changes in the external world. He will put these changes on the level of ontology and think that these are inevitable, such as the blessings and failures of the world. This ontological understanding of social changes has not given Zhuangzi the motivation to improve society and develop society in a better direction.

References

1. Robert D. Richardson Jr., *Emerson: The Mind on Fire* (University of California Press, 1995).
2. Ralph Waldo Emerson, *The Best Read Naturalist*, ed. Michael P. Branch and Clinton Mohs (University of Virginia Press, 2017).
3. Ralph Waldo Emerson, *The Portable Emerson*, ed. Jeffrey S. Cramer (Penguin Books, 1946).
4. Immanuel Kant, *Critique of Pure Reason*, trans. Marcus Weigelt (Cambridge University Press, 2007).
5. Ralph Waldo Emerson, *The Conduct of Life* (The Riverside Press, 1860).
6. Emerson: *Essays & Poems*, ed. Joel Porte (Literary Classics of the United States 1983).
7. Peiyuan Meng, *Emotion and Reason* (Beijing: China Renmin University Press, 2009).
8. *A Source Book in Chinese Philosophy*, trans. Wing-Tsit Chan (Princeton University Press, 1963).
9. Zhuangzi, *Zhuangzi: The Essential Writings*, trans. Brook Ziporyn (Hackett Publishing Company, 2009).
10. Ralph Waldo Emerson, *Essays and English Traits* (New York: P. F. Collier, 1909).
11. Ralph Waldo Emerson: *English Traits & Representative Men & Other Essays*, ed. Ernest Rhys (London: J. M. Dent & Sons, Ltd, 1908).
12. Confucius, *The Analects*, trans. D. C. Lau (Penguin Classics, 1979).
13. Mencius, *Mencius*, trans. D. C. Lau (Penguin Books, 2004)

Chapter 10
The False Self

The false self refers to the self that deviates from virtue. It is the self which must be overcome so that we can achieve the true self, which is virtuous self. In both Emerson and Chinese philosophy, some special terms are used to convey the idea of overcoming the false self. Emerson's "surrender", Confucius' "overcome" (克), and Zhuangzi's "lose" (丧) all convey this idea of overcoming the false self. In terms of the false self in Emerson and Chinese philosophy, one point needs to be stressed here is that the false self is not evil. In this aspect, it differs from many Western philosophers, who often use evil to express the false self. For example, Aristotle's concept of moderation as an important virtue is related to human beings' tendency to go to extremes, which is considered as evils by him. Kant argues that there is a fundamental evil in human nature, which is ineradicable. Kant also reinterpreted the Christian doctrine of sin in terms of radical evil. This evil is not manifested as a natural tendency of humans, but refers to a reversal of order, mainly referring to the pursuit of happiness above the pursuit of moral laws. Emerson and Chinese philosophy are consistent on this point: that is, they all believe that the false self that deviates from virtue is not evil, and there is no evil in human nature. On the contrary, human nature is all good, although they have many differences in interpreting this goodness.

1 The False Self in Emerson

Emerson employs many different terms to express the idea of the false self. These terms include egotism, individualism, and self-willed. I will discuss Emerson's understanding of the false self from three aspects: (1) Living around the welfare of the individuals, without pursuing the ultimate values of truth, goodness, and beauty; (2) Living according to opinion and conflating it with one's personal will; (3) Worshipping results and losing the close relationship with nature. One thing I would like to

S. Gao, *Comparative Studies Between Emerson and Chinese Philosophy*, Nature Aesthetics: Interdisciplinary and Intercultural Dialogue,
https://doi.org/10.1007/978-981-95-0421-3_10

stress here is that Emerson doesn't think the above types of the false self is evil. It can be seen from the following quote, "Saints are sad, because they behold sin (even when they speculate) from the point of view of the conscience, and not of the intellect; a confusion of thought. Sin, seen from the thought, is a diminution, or *less*; seen from the conscience or will, it is pravity or *bad*. The intellect names it shade, absence of light, and no essence. The conscience must feel it as essence, essential evil. This it is not; it has an objective existence, but no subjective."[1] For him, evil has no essence and it is just deviation from the goodness.

First, living around the individual benefits. This type of self refers to the idea that a person is only concerned with his or her own individual interests and welfare, living around individual joys and sorrows. Emerson thinks that this type of self has stripped the connection between the individual and the whole, which is false. He writes, "So deep is our sense of the necessity of resting on nature and of universal motives, that, we call such, real men;—whilst men acting for by—ends, or not from central and constitutional reason, false and superficial. This acting after your constitution and for that which is always the same, as, justice, goodness, human freedom, and benefit—we call reality."[2] A virtuous man, in Emerson's view, is one who has realized the true self. Such a person relies on natural and universal motives when acting, which are motives for justice, goodness, human freedom, and welfare. The false self, on the other hand, pursues other motives. Emerson sees this kind of self as a type of individualism. This individualism will immerse people in their narrow individual world and cause a lot of worries. In Emerson's view, when people act from ultimate values such as justice, truth, or beauty, the power of nature will actively join in and empower a person.

Second, Emerson's concept of self-willed refers to the idea that the person lives by opinion. Emerson elaborates in the following passage, "God offers to every mind its choice between truth and repose. Take which you please,—you can never have both. Between these, as a pendulum, man oscillates. He in whom the love of repose predominates will accept the first creed, the first philosophy, the first political party he meets,—most likely his father's. He gets rest, commodity, and reputation; but he shuts the door of truth."[3] Through this passage, we can see that the false self that Emerson is talking about mainly refers to the self that chooses not to move forward, but to be content with tranquility. This tranquility comes from living according to "perfect" opinions, not according to truth. The "perfect" opinions mentioned by Emerson are relative to imperfect opinions. He refers to the values that people accept in daily life without reflection, including the first principles, the first philosophy, and the first political party, etc., but these are not the truth. Emerson emphasizes that scholars need "self-denial", which is to deny these lives based on imperfect opinions.

---

[1] Ralph Waldo Emerson, *The Portable Emerson*, ed. Jeffrey S. Cramer (Penguin Books, 2014), p. 275.

[2] Ralph Waldo Emerson, *The Later Lectures of Ralph Waldo Emerson*, ed. Ronald A. Bosco and Joel Myerson (Athens: University of Georgia Press, 2001), p. 34.

[3] Emerson: *Essays and Lectures*, ed. Joel Porte (Literary Classics of the United States, 1983), pp. 425–426.

The false self as understood by Emerson is also connected to a kind of power, which can bring worldly success, but it is far from the truth and needs to be restrained. He writes, "Our American people cannot be taxed with slowness in performance or in praising their performance. The earth is shaken by our engineries. We are feeling our youth and nerve and bone. We have the power of territory and of seacoast, and know the use of these. ... Our political constitution is the hope of the world, and we value ourselves on all these feats."[4] It can be seen that the popular standards of success in Emerson's lifetime include political expansion and the technical success obtained by promoting industrialization. These practices also require great power, but Emerson believes that these powers deviate from the fundamental moral power of being human.

Third, the false self also refers to worshipping the results, which will cause alienation from appreciation of the beauty of nature. The results as understood by Emerson are relative to the cause and are the results contained in the cause. This cause refers to the ultimate cause, that is, God. The results include: (1) various institutions on the earth, such as the church, etc; (2) humans and all things in nature; (3) various political parties formed in society. Emerson calls people's worship of results Idolatry. Emerson writes:

> There are but two things, or but one thing & its shadow—Cause & Effect, and Effect is itself worthless if separated from Cause. It is Cause still that must be worshipped in Effect; so that it is only one thing. The worship of Effect is Idolatry. The Church including under this name, Doctrine, Forms, Discipline, Members, is the instant Effect: weak man adheres to the Effect & lets God go... The indisposition of men to go back to the source & mix with Deity is the reason of degradation & decay. Education is expended in the measurement & imitation of Effects: in the study of Shakespeare for example as itself a perfect being is with every scholar. Thus the College becomes idolatrous—a temple full of idols. Shakespeare will never be made by the study of Shakespeare. I know not how directions for greatness can be given. Yet greatness may be inspired.[5]

Emerson characterizes the worship of results as a form of idolatry. This idol could be a specific person, an institution, nature, and so on. Emerson argues that the emergence of these phenomena can only be explained by reference to an ultimate cause, which manifests human creativity. However, this creativity is itself inspired by the same ultimate cause. For him, when we fail to recognize this ultimate cause and instead worship its products, we distance ourselves from God. This is also the false self. Emerson's reverence for the ultimate cause is very similar to Spinoza. In Spinoza's view, we have a duty to understand the ultimate cause, which is God. He opposes the love of specific things in this world, including the pursuit of love, power, and wealth, etc. He thinks that if we give our deepest love to these specific things, people will eventually suffer. But if we combine our soul's cognition with the ultimate cause, which is God, then we will achieve the ultimate good. Emerson also emphasizes our cognition and love of God, but the biggest difference between Emerson and Spinoza

[4] Ralph Waldo Emerson, *Society and Solitude: Twelve Chapters in The Complete Works of Ralph Waldo Emerson*, vol. VII, ed. Edward Waldo Emerson (Boston and New York: Houghton Mifflin Company, 1904), p. 283.

[5] Joseph Urbas, *The Philosophy of Ralph Waldo Emerson* (Routledge, 2021), p. 193.

is that Emerson emphasizes the importance of knowing nature and the world in experience, which is the basis for knowing the ultimate entity, which is God. But Spinoza believes that the knowledge of nature in experience is inferior knowledge, while the intuitive knowledge of God is the highest knowledge and truth.

According to Emerson, when a person worships the result, he distances himself from nature. People will spend more time in the church and ignore the original relationship between man and nature. Emerson regards nature as the standard by which to judge whether a person has fallen. If a person cannot hear the birds singing, cannot appreciate the beauty of nature, this self is also false. He writes, "As we degenerate, the contrast between us and our house is more evident. We are as much strangers in nature as we are aliens from God. We do not understand the notes of birds. The fox and the deer run away from us; the bear and tiger rend us. We do not know the uses of more than a few plants, as corn and the apple, the potato and the vine. Is not the landscape, every glimpse of which hath a grandeur, a face of him? Yet this may show us what discord is between man and nature, for you cannot freely admire a noble landscape if laborers are digging in the field hard by. The poet finds something ridiculous in his delight until he is out of the sight of men."[6] It can be seen that when a person falls, he becomes alienated from nature. The specific manifestation of alienation from nature is not understanding the language of nature, including the chirping of birds and the uses of plants, etc. Emerson regards the inability of people to freely appreciate the beauty of nature as a manifestation of disharmony between man and nature, and also as a manifestation of the false self.

2 The False Self in Confucianism

Confucianism often uses terms such as desire (欲), small man (小人), and egoism (私) to express the idea of the false self. The false self in Confucianism is characterized by three features: (1) Narrow self. It refers to the pursuit of desire for profit, fame or personal interests; (2) Lack of respect for the rituals (礼); (3) Worshipping God. In Confucianism, God refers to transcendental powers which are beyond human beings' control.

First, Confucianism frequently uses the term "small man" (小人) to characterize the narrow self. In *The Analects* and *Mencius,* many passages are related to small man.

> The Master said, 'While the gentleman cherishes benign rule, the small man cherishes his native land. While the gentleman cherishes a respect for the law, the small man cherishes generous treatment.'[7]
>
> Tzu-chang asked, 'What must a Gentleman be like before he can be said to have got through?' The Master said, 'What on Earth do you mean by getting through?' Tzu-chang answered,

---

[6] Ralph Waldo Emerson, *The Portable Emerson,* ed. Jeffrey S. Cramer, (Penguin Books, 2014), p. 48.

[7] Confucius, *The Analects*, trans. D. C. Lau, (Penguin Books, 1979), p. 73.

> 'What I have in mind is a man who is sure to be known whether he serves in a state or in a noble family.' The Master said, 'That is being known, not getting through. Now the term "getting through" describes a man who is straight by nature and fond of what is right, sensitive to other people's words and observant of the expression on their faces, and always mindful of being modest. Such a man is bound to get through whether he serves in a state or in a noble family…'[8]

> The Master said, 'If one is guided by profit in one's actions, one will incur much ill will.'[9]

> The parts of the person differ in value and importance. Never harm the parts of greater importance for the sake of those smaller importance, or the more valuable for the sake of the less valuable. He who nurtures the parts of smaller importance is a small man; he who nurtures the parts of greater importance is a great man.[10]

> The organs of hearing and sight are unable to think and can be misled by external things. When one thing acts on another, all it does is to attract it. The organ of the heart can think. But it will find the answer only if it does think; otherwise, it will not find the answer. This is what Heaven has given me.[11]

In *The Analects* and *Mencius*, the term of small man often appears together with the term of great man. A small man doesn't pursue virtue rather than narrow self-interests, which include the sensual pleasures, profit, and reputation. For Confucianism, a small man's pursuit of these things deviates from the pursuit of virtue. But one thing I would like to stress here is that Confucianism shares a similar view with Emerson: If we don't pursue virtue which represents the true self, our self will be false. But the self is not evil. Human nature for them is originally good, which has its source in God or Dao, and evil doesn't exist from the metaphysical dimension.

Second, lack of respect for rituals. Confucianism respects rituals that have been handed down from the past to govern human relationships. Confucianism would not challenge this. A passage about rites from *The Analects* well illustrates this point: "Yu Tzu said, 'Of the things brought about by the rites, harmony is the most valuable. Of the ways of the Former kings, this is the most beautiful, and is followed alike in matters great and small, yet this will not always work: to aim always at harmony without regulating it by the rites simply because one knows only about harmony will not, in fact, work.'"[12] In Confucianism, the main function of rituals is to achieve harmony among the people. Harmony is based on virtue, and it is also considered as being beautiful. Virtue is the foundation for this beauty. Emerson also stresses the importance of history and previous generations in shaping ourselves. But he differs from Confucianism that Emerson stresses progress in morals and development of cognitive abilities of human beings. For him, all virtues and values will be transcended by new ones. Emerson strongly criticized ceremonies or rituals in a church to connect us to God. For him, nature is the best interpreter of God, and everyone has the ability to have a direct connection with God.

---

[8] Ibid., p. 116.

[9] Ibid., p. 73.

[10] Mencius, *Mencius*, trans. D. C. Lau (Penguin Books, 1970), p. 168.

[11] Ibid.

[12] Confucius, *The Analects*, trans. D. C. Lau (Penguin Books, 1979), p. 61.

Third, worshipping God. For Confucianism, if we worship God, our self will be false. In *The Analects*, a few passages are related to his attitude towards God. "Fan Ch'ih asked about wisdom. The Master said, 'To work for the things the common people have a right to and to keep one's distance from the gods and spirits while showing them reverence can be called wisdom.'";[13] "The topics the Master did not speak of were prodigies, force, disorder and gods."[14] By God, Confucius refers to transcendental powers which we will never be able to understand. For him, the only world which has metaphysical importance is the world we live in everyday life. We should immerse ourselves in this world and get to know it deeply within this life. Emerson stresses just the opposite. For him, the transcendental world, which is the world of Oversoul, or God has the ultimate reality and it is the original source of our everyday life. We should know God first.

3 The False Self in Daoism

Within Confucian thought, the true self is realized through the continuous cultivation of moral virtues, particularly the four cardinal virtues of benevolence, righteousness, propriety, and wisdom. In stark contrast, Daoism, particularly as articulated in the Zhuangzi, views these virtues as detrimental to the authentic self, critiquing them as artificial constraints on natural vitality. Before I address Daoism's understanding of the false self, I will first examine Zhuangzi's critique of Confucian virtues of benevolence and righteousness before exploring Daoism's core categories of the false self which are desire, formed mind and mechanical mind.

Zhuangzi gives a strong criticism towards Confucian moral ideas from three aspects: (1) Fostering fake emotions. Zhuangzi perceives Confucian benevolence and righteousness as fostering fake emotions, a theme vividly explored in the "Yufu" chapter. He points out Confucian etiquette mandates ritual displays, such as crying at funerals, even when genuine grief is absent. He writes, "He who lacks purity and sincerity cannot move others. Therefore he who forces himself to lament, though he may sound sad, will awaken no grief. He who forces himself to be angry, though he may sound fierce, will arouse no awe."[15] Zhuangzi provides a sharp criticism towards Confucianism in terms of the fake emotions people tend to develop under the pressure of Confucian pursuit of moral virtues. He provides a new concept, which is zhen. Zhuangzi's idea of truth differs from Western concept of truth greatly. For Zhuangzi, truth refers to the intersubjective and authentic feelings among the people. (2) Detrimental to vitality. Zhuangzi contends that the moral judgments imposed by benevolence and righteousness stifle individual vitality. By prioritizing these virtues, society encourages their pursuit at the expense of ontological concern

[13] Ibid., p. 84

[14] Ibid., p. 88.

[15] Zhuangzi, *The Complete Works of Chuang Tzu*, trans. Burton Watson, (New York: Columbia University Press, 1968), p. 349.

for life itself, a critique implicit in his broader philosophy. This external focus, he suggests, drains the spontaneous energy of the self. (3) The fading of true love. Zhuangzi asserts that Confucian virtues erode authentic filial piety and love, replacing them with performative morality." Only by relinquishing these ideals, he argues, can natural affection flourish. In critique of the Confucian concept of the true self and its pernicious influence on a person's vitality, Zhuangzi puts forward three key concepts which represents Zhuangzi's understanding of the false self. These concepts are desire, formed mind (成心), and mechanical mind (机心).

First, desire is one aspect of the false self according to Zhuangzi. This idea can be seen from the following passages. "Deep in their passions and desires, they are shallow in the workings of Heaven."[16]; "Eminence and wealth, recognition and authority, fame and profit—these six are the delusions of the will."[17] For Zhuangzi, the desire is often connected to wealth, recognition, fame, political power, and knowledge. For him, desire will harm the virtue of human beings. Virtue in Zhuangzi is deeply related to vitality and intuition of highest reality, Dao.

Second, the concept of formed mind also represents Zhuangzi's understanding of the false self. He mentions the idea of formed mind in the following passage, "If a man follows the mind given him and makes it his teacher, then who can be without a teacher? Why must you comprehend the process of change and form your mind on that basis before you can have a teacher? Even an idiot has his teacher. But to fail to abide by this mind and still insist upon your rights and wrongs—this is like saying that you set off for Yüeh today and got there yesterday"[18] For Zhuangzi, the formed mind is the false self. By formed mind, he means that we conduct the moral judgment by using our own standard without following Dao. This false self will cause the emotional turmoil which is vividly expressed by Zhuangzi in the following passage, "Day after day they use their minds in strife, sometimes grandiose, sometimes sly, sometimes petty. Their little fears are mean and trembly; their great fears are stunned and overwhelming. They bound off like an arrow or a crossbow pellet, certain that they are the arbiters of right and wrong. They cling to their position as though they had sworn before the gods, sure that they are holding on to victory. They fade like fall and winter—such is the way they dwindle day by day."[19] This passage illustrates the emotional turmoil caused by the false self. However, I would like to stress that Zhuangzi's idea of the false self doesn't have the meaning of evil. Kim-chong Chong in his article "Zhuangzi and the Issue of Human Nature"[20] well examined this topic. He refutes A. C. Graham's claim that Zhuangzi sees human nature as "self-evidently bad", arguing it's a misreading of such passages. Formed mind for Zhuangzi is one type of the false self but it is not evil. It is the self which deviates from the virtue.

---

[16] Ibid., p. 78.

[17] Ibid., p. 259.

[18] Ibid., pp. 38-39.

[19] Ibid., p. 37.

[20] Kim-chong Chong, "Zhuangzi and the Issue of Human Nature", Dao (2023) 22:237-254.

Third, Zhuangzi employs the concept of mechanical heart to covey the idea of the false self. Zhuangzi mentions the mechanical heart in the this passage, "With a machine heart in your breast, you've spoiled what was pure and simple; and without the pure and simple, the life of the spirit knows no rest. Where the life of the spirit knows no rest, the Way will cease to buoy you up. It's not that I don't know about your machine—I would be shamed to use it! '"[21] For Zhuangzi, if we try to use technology to improve our daily working on the land, our nature will be spoiled. For him, body plays an important role in philosophy. Farming activities without technology can develop our inborn nature.

Confucianism's true self, which is rooted in benevolence and righteousness becomes the false self in Zhuangzi, a construct that Zhuangzi deems harmful to human virtue. From the above, we can see that the true self as understood by Confucianism becomes the false self. He asserts, "My definition of expertness has nothing to do with benevolence and righteousness; it means being expert in regard to your Virtue, that is all. My definition of expertness has nothing to do with benevolence or righteousness, it means following the true form of your inborn nature, that is all."[22] Zhuangzi's conception of virtue stresses our relationship with the Dao. If we follow the Dao, our inborn nature will be developed and we will attain our virtue. For him, the Confucian virtue of benevolence and righteousness will harm our inborn nature since it is deviated from the Dao.

References

1. Ralph Waldo Emerson, *The Portable Emerson*, ed. Jeffrey S. Cramer (Penguin Books, 2014).
2. *The Later Lectures of Ralph Waldo Emerson*, ed. Ronald A. Bosco and Joel Myerson (University of Georgia Press).
3. Ralph Waldo Emerson, *Essays and Lectures*, ed. Joel Porte (Literary Classics of the United States, 1983).
4. Ralph Waldo Emerson, *Society and Solitude* (twelve chapters), in *Complete Works of Ralph Waldo Emerson*, ed. Edward Waldo Emerson (Boston and New York: Houghton Mifflin Company, 1904).
5. Joseph Urbas, *The Philosophy of Ralph Waldo Emerson* (Routledge, 2021).
6. Confucius, *The Analects*, trans. D. C. Lau (Penguin Books, 1979).
7. Mencius, *Mencius*, trans. D. C. Lau (Penguin Books, 1970).
8. Zhuangzi, *The Complete Works of Chuang Tzu*, trans. Burton Watson (Columbia University Press, 1968).
9. Kim-chong Chong, "Zhuangzi and the Issue of Human Nature," *Dao* (2023) 22.

[21] Zhuangzi, *The Complete Works of Chuang Tzu*, Burton Watson (New York and London: Columbia University Press, 1968), p. 134.

[22] Ibid., p. 103.

Chapter 11
Virtue and Self-Transcendence

The spirituality in Emerson and Chinese philosophy has similar characteristics. For them, they both think that virtue represents the very essence of human beings. Virtue contains the effort of overcoming the false self to achieve the true self, which is to live a life in harmony with nature. In part II of this book, I have examined the value dimension of nature. It is the metaphysical foundation for self-transcendence towards the value of goodness. In terms of the relationship between virtue and self-transcendence, Jennifer A. Frey and Candace Vogler have devoted an entire book titled *Self-Transcendence and Virtue: Perspectives from Philosophy, Psychology, and Theology* to this topic. Their approach is interdisciplinary which contains perspectives from philosophy, psychology, and theology. Their devotion to the relationship between virtue and self-transcendence is based on the consensus among the scholars from these three disciplines. The general idea is that virtuous life is a meaningful life which includes cultivation of an ability to transcend the egoism to embrace the larger whole. In this chapter, I will focus on philosophical perspectives on the relationship between virtue and self-transcendence. In Emerson and Chinese philosophy, virtue is in essence to overcome the false self to transcend towards the ultimate value of goodness. In Emerson and Chinese philosophy, God and Dao are the bearers of the ultimate value of Good respectively. For them, we have moral obligation to achieve self-transcendence.

1 Self-transcendence: the Essence of Virtue

In Emerson and Chinese philosophy, virtue's relationship with self-transcendence can be seen from their ways of expressions of virtue. The concept of virtue in the works of Emerson and Chinese philosophy is frequently interpreted in terms of three key notions: transformation, tendency, and change. All of these words imply the idea of self-transcendence. Let us begin with Emerson's perspective on virtue. He writes,

S. Gao, *Comparative Studies Between Emerson and Chinese Philosophy*, Nature Aesthetics: Interdisciplinary and Intercultural Dialogue,
https://doi.org/10.1007/978-981-95-0421-3_11

"There is no virtue which is final; all are initial. The virtues of society are vices of the saint. The terror of reform is the discovery that we must cast away our virtues, or what we have always esteemed such, into the same pit that has consumed our grosser vices:—'Forgive his crimes, forgive his virtues too, Those smaller faults, half converts to the right.'"[1] For Emerson, no virtue is static or complete since all of the virtues within a certain society or historical time will be transcended. What society currently celebrates as virtuous will soon be replaced by higher ideals. Virtue, in his view, is dynamic and perpetually evolving. This idea is best expressed in the following passage, "The epochs of our life are not in the visible facts of our choice of a calling, our marriage, our acquisition of an office, and the like, but in a silent thought by the wayside as we walk; in a thought which revises our entire manner of life and says, —'Thus hast thou done, but it were better thus.' And all our after years, like menials, serve and wait on this, and according to their ability execute its will. This revisal or correction is a constant force, which, as a tendency, reaches through our lifetime."[2] Here, Emerson portrays virtue as an ongoing process of transcending the false self, guided by an inner thought that illuminates a better way. This transformation is not driven solely by personal will but by a spiritual law that flows through the individual, reshaping their existence when they surrender to its guidance. From the Chapter 9 of this book, we know that Emerson's understanding of virtue is closely related to the cognitive abilities of the mind of human beings. Self-transcendence for him refers to the elevation of cognitive abilities. He writes, "The soul's advances are not made by graduation, such as can be represented by motion in a straight line, but rather by ascension of state, such as can be represented by metamorphosis, —from the egg to the worm, from the worm to the fly." [3] For him, the soul's metamorphosis is the development of cognition. It is also the development of virtues.

Chinese philosophy similarly emphasizes transformation in its interpretation of virtue. The Chinese term for virtue, *de*, as explained in *Shuowen Jiezi*, China's oldest dictionary, carries the connotation of "rising" or self-elevation, suggesting a movement toward the Dao (the Way). Confucius encapsulates this idea of self-transcendence in his teaching: "Yen Yüan asked about benevolence. The Master said, 'To return to the observance of the rites through overcoming the self constitutes benevolence. If for a single day a man could return to the observance of the rites through overcoming himself, then the whole Empire would consider benevolence to be his. However, the practice of benevolence depends on oneself alone, and not on others."[4] This passage distinguishes two selves: the virtuous self, which realizes benevolence, and the false self, which must be restrained because it deviates from propriety. Likewise, Zhuangzi expresses the idea of self-transcendence with the phrase "I have lost me". He writes, "Ziqi of the Southern Wall was reclining against a low table on the ground, releasing his breath into Heaven above, all in a scatter, as

[1] Emerson, *The Portable Emerson*, ed. Jeffery S. Cramer, (Penguin Books, 2014), p. 234.

[2] Ibid., p. 206.

[3] Ibid., p. 214.

[4] Confucius, *The Analects*, trans. D. C. Lau (Penguin Books, 1979), p. 112.

if loosed from a partner. Yancheng Ziyou stood in attendance before him. 'What has happened here?' He said. 'Can the body really be made like dried wood, the mind like dead ashes? What reclines against this table now is not what reclined against it before.' Ziqi said, 'A good question, Yan! What has happened here is simply that *I* have lost me'".[5] In this dialogue, "*I* have lost me" refers to shedding the false self—the ego bound by worldly attachments—to reveal the true self, united with the Dao. A key distinction emerges here: While Chinese philosophy emphasizes the active role of human will in this gradual process, Emerson views transformation as an immediate act facilitated by a divine spiritual force from God.

For Emerson, self-transcendence moves us from the false self toward the true self and it culminates in living in harmony with nature. The concept of harmony is an alternative expression of the ultimate value of truth, goodness, and beauty. Chinese philosophy shares the similar vision in terms of the life in harmony with nature and value of goodness, though their conceptions differ significantly. Emerson writes, "There is deep reason for the love of nature that has characterized the highest minds. The soul and the body of things are harmonized; therefore the deeper is a man's insight into the spiritual laws the more intense will be his love of the works of nature."[6] For Emerson, there is a correspondence between the soul and the material nature. The more one grasps this law, the greater their appreciation for nature's beauty becomes. He writes, "We are adapted to infinity. We are hard to please, and love nothing which ends: and in nature is no end; but every thing, at the end of one use, is lifted into a superior, and the ascent of these things climbs into dæmonic and celestial natures. Creative force, like a musical composer, goes on unweariedly repeating a simple air or theme, now high, now low, in solo, in chorus, ten thousand times reverberated, till it fills earth and heaven with the chant."[7] Nature, in Emerson's view, embodies an infinite creative process, reflecting the divine power of creation. By aligning with nature, we unite with this boundless force.

In contrast, Chinese philosophy frames harmony between humanity and nature through the lens of moral unity with Heaven and Earth. In Confucianism, it is evident in Mencius's teaching: "The gentleman's attitude toward things is to love them but not to be benevolent; toward the people, to be benevolent but not to be close. To be close to one's relatives, benevolent to the people, and loving toward things."[8] Here, "things" include natural objects, but Confucian love originates in familial bonds and weakens as it extends outward to nature. For Daoism, Zhuangzi offers a different vision: "Emptiness, stillness, limpidity, silence, inaction—these are the level of Heaven and Earth, the substance of the Way and its Virtue. Therefore the emperor, the king, the sage rest in them. Resting, they may be empty; empty, they may be full; and fullness

---

[5] Zhuangzi, *Zhuangzi: The Essential Writings with Selections from Traditional Commentaries*, trans. Brook Ziporyn (Hackett Publishing, Indianapolis, 2009), p. 9.

[6] *The Best Read Naturalist*, eds. Michael P. Branch and Clinton Mohs (University of Virginia Press, 2017), p. 61.

[7] Ralph Waldo Emerson, *Essays and Lectures*, ed. Joel Porte (Literary Classics of the United States, 1983), p. 669.

[8] Mencius, *Mengzi: With Selections from Traditional Commentaries*, trans. Bryan W. Van Norden, (Hackett Publishing, Indianapolis, 2008), p. 183.

is completion. Empty, they may be still; still, they may move; moving, they may acquire. Still, they may rest inaction; resting in inaction, they may demand success from those who are charged with activities. Resting in inaction, they may be merry; being merry, they may shun the place of care and anxiety, and the years of their life will be long."[9] For Zhuangzi, harmony arises when the mind empties itself of the false self, aligning with the virtue of inaction of Heaven and Earth. Unlike Emerson's focus on nature's creative infinity, Chinese philosophy projects human ethical values onto the cosmos, using this framework to guide interpersonal relationships.

2 Self-transcendence: Trinity and Unity

In the above, I mentioned that Emerson and Chinese philosophy both pursue a life of living in harmony with nature. In Emerson, living in harmony with nature is based on the metaphysical concept of trinity. Emerson reinterprets the traditional Christian trinity. In the new trinity, nature becomes the medium for the relationship between human beings and God. In Chinese philosophy, unity between human beings and nature is the foundation for a life in harmony with nature. Emerson's concept of trinity promotes human beings' understanding and appreciation of nature. In contrast, Chinese philosophy, unity between human beings and nature is human-centered and it encourages us to know more about human beings, although it doesn't neglect understanding of nature.

In the life of living in harmony with nature, Emerson puts emphasis on knowing the truth of nature since it is the medium for human beings to know God. For Emerson, we have moral obligation to know the truth in nature. He writes, "But if the constructive powers are rare, and it is given to few men to be poets, yet every man is a receiver of this descending holy ghost, and may well study the laws of its influx. Exactly parallel is the whole rule of intellectual duty to the rule of moral duty. A self-denial, no less austere than the saint's, is demanded of the scholar. He must worship truth, and forego all things for that, and choose defeat and pain, so that his treasure in thought is thereby augmented."[10] For Emerson, knowing the truth is both a rational and moral imperative, intertwining intellectual and ethical duties. In contrast, Confucianism prioritizes moral obligations within specific interpersonal relationships. The persons in Chinese philosophy are not the abstract and rational persons as interpreted by Kant, but concrete ones embedded in the social relationships. And the most important relationship is our relationship with parents. Confucianism stresses our primary moral duty towards parents. This idea is well illustrated in the following passages from *The Analects.*

[9] Zhuangzi, *The Complete Works of Chuang Tzu*, trans. Burton Watson (New York and London: Columbia University Press, 1968), p. 142.

[10] Ralph Waldo Emerson, *Essays and Lectures*, ed. Joel Porte (The Library of America, 1983), p. 425.

> Yu Tzu said, '...The gentleman devotes his efforts to the roots, for once the roots are established, the Way will grow therefrom. Being good as a son and obedient as a young man is, perhaps, the root of a man's character.'[11]

> The Master said, 'Observe what a man has in mind to do when his father is living, and then observe what he does when his father is dead. If, for three years, he makes no changes to his father's ways, he can be said to be a good son.'[12]

> Tzu-yu asked about being filial. The Master said, 'Nowadays for a man to be filial means no more than that he is able to provide his parents with food. Even hounds and horses are, in some way, provided with food. If a man shows no reverence, where is the difference?'[13]

Emerson stresses our duty to knowing the truth, while Confucianism focuses on our primary duty towards parents and the people who are in close relationships. These duties towards our parents are meticulously prescribed. Youzi's assertion that filial piety and fraternity form the foundation of benevolence underscores this priority. Filial devotion entails adhering to propriety, maintaining a respectful demeanor, and honoring parental will. Beyond parents, obligations extend to siblings and friends within blood or intimate ties. Zengzi's famous "threefold daily self-examination" (吾日三省吾身), for instance, includes duties to friends, expressed through daily care. Confucianism also advocates extending these obligations beyond close kin, though their intensity diminishes with relational distance.

In his essay "Self-Reliance", Emerson critiques socially prescribed duties and rejects the conventional moral obligations attached to social roles such as father, husband, or citizen. Instead, he advocates duties grounded in personal intuition or genius rather than social convention. Emerson proclaims, "I shall endeavor to nourish my parents, to support my family, to be the chaste husband of one wife,— but these relations I must fill after a new and unprecedented way. I appeal from your customs. I must be myself. I cannot break myself any longer for you, or you."[14] For Emerson, the supreme obligation lies not in blood ties but in fidelity to the eternal law—the divine or natural force. Discerning this law takes precedence, and cultivating one's cognitive capacity is essential to this duty, redirecting his emphasis toward intellectual self-improvement. A question arises here: why does Confucianism stress our primary duty towards our parents instead of knowing the truth of God? The answer lies in their different conceptions of life. For Emerson, the transcendental realm transcends the physical world, serving as its cause. In Confucianism, life comes from the interaction of *yin* and *yang* energies. Parents, who are the combination of male and female, represent the interaction of opposing energies. It has the metaphysical importance.

Daoism, in contrast, focuses on an obligation to life itself—a vitality rooted in material *qi* infused with spirituality. When *qi* flows harmoniously, life flourishes, awakening an intuitive cognition Daoism calls "spirit" (shen). Daoism rejects Confucian moral virtues of benevolence and righteousness. For Daoism, these virtues will

---

[11] Confucius, *The Analects,* trans. D.C. Lau (Penguin Books, 1979), p. 59.

[12] Ibid., pp. 60–61.

[13] Ibid., p. 64.

[14] Ralph Waldo Emerson, *The Portable Emerson*, ed. Jeffrey S. Cramer (Penguin Books, 2014), p. 164.

diminish the vitality of body and mind. Zhuangzi stresses nurturing life—not merely to prolong it, but to ignite its vitality. The following quotes from Zhuangzi can help us better understand this point. In the essay titled "The Primacy of Nourishing Life," he cautions, "The flow of my life is always channeled by its own boundaries, but the mind ben on knowledge never is. A flow channeled by its own boundaries is endangered when forced to follow something that is not, and trying then to rescue it with the doings of the knowing mind only makes the danger worse."[15] In "Fathoming Life", he explains, "Why then is it worthwhile to let go of your concerns and forget all about life? Because when concerns are let go of, the body is no longer labored. When life is forgotten, the seminal quintessence of vitality remains undamaged. When the body is intact and the seminal quintessence of vitality restored, you are one with Heaven. Heaven and earth are the parents of all things."[16] By "forgetting life", Zhuangzi refers to the pursuit of all types of happiness including long life, fame, knowledge, and so on. For Zhuangzi, our body will be full of vitality if we forget about life. Vitality is the important value for Zhuangzi. This perspective sharply opposes Emerson's, who extols the duty to pursue truth, whereas Zhuangzi views objective knowledge as elusive and subordinate to life's dynamism.

3 Self-transcendence: Moral Cultivation

Both Emerson and Chinese Philosophers stress that attainment of self-transcendence needs moral cultivation. In terms of moral cultivation, Emerson stresses the top-down power from God while Confucianism and Daoism both stress the power of our will in moral cultivation.

Emerson's concept of moral sentiment can help us better understand the top-down power's role in moral cultivation. Emerson writes, "All things are moral. That soul which within us is a sentiment, outside of us is a law. We feel its inspiration; but in history we can see its fatal strength."[17] Sentiment in this passage doesn't refer to the moral emotions of human beings. Rather, it refers to the moral intuition of human beings which originates from God. By external soul, he refers to the impersonal God which is the combination of ultimate values of truth, goodness, and beauty. This sudden awakening is immediate, which is similar to enlightenment in Zen Buddhism. For Emerson, the human mind's rational faculty seeks absolute truth, serving as a conduit for divine creativity. This gateway to God remains open and it will never be closed. God will actively reveals itself to the mind. However, we must transcend the false self to be able to receive God's inspiration.

---

[15] Zhuangzi, *Zhuangzi: The Essential Writings with Selections from Traditional Commentaries*, trans. Brook Ziporyn (Hackett Publishing Company), p. 21.

[16] Ibid., p. 77.

[17] Ralph Waldo Emerson, *The Portable Emerson*, ed. Jeffrey S. Cramer (Penguin Books 2014), p. 178.

In contrast, Confucianism stresses our own will power in moral cultivation. One passage in *The Analects* can better illustrate this idea, "The Master said, 'It is Man who is capable of broadening the Way. It is not the Way that is capable of broadening Man.'"[18] In Confucianism, the Way cannot give people transcendental power to practice virtue. The attainment of virtue needs sustained human effort and it is a very long process. Confucius expresses this process concisely, "The Master said, 'At fifteen I set my heart on learning; at thirty I took my stand; at forty I came to be free from doubts; at fifty I understood the Decree of Heaven; at sixty my ear was attuned; at seventy I followed my heart's desire without overstepping the line."[19] This progression underscores the deliberate and lifelong cultivation of virtue through human will, devoid of the spontaneous divine infusion Emerson envisions. Confucius himself also went through decades of effort to gradually achieve the highest state of virtue, that is, to know the will of Heaven, and to completely unify the observance of rites and the inner pursuit of virtue. But Confucius also laments that among the people he knew, few put the pursuit of virtue first. It can be seen that people need to struggle with many desires to achieve moral obligations.

Daoism, too, underscores the subjective effort of the will in cultivating virtue. This idea is best demonstrated in the following two dialogues in Zhuangzi. The first dialogue takes place between Nanbo Zikui and Lady Ju. Nanbo Zikui about ways for him to teach the Dao to Puliang Yi who has the innate powers of a sage. He says, "After I kept at him, with it for three days, he could oust the entire world from himself. After seven days more, he could likewise oust all *things,* and after another seven, even *life.* Only when life was put outside himself could the dawn break through him. Only with the breakthrough of dawn could he see the Singularity. After seeing the Singularity, he could free himself of past and present."[20]

Similarly, a dialogue between Yanhui and Confucius in Zhuangzi can also illustrate the long process of inner effort in the process of forgetting. When Yanhui told Confucius that he was making progress in forgetting. The process includes three steps and it takes many days. (1) forgetting humanity and responsibility; (2) forgetting ritual and music; (3) just sitting there and forget. For the first two steps, Confucius points out that although these are good efforts but still cannot be counted as real forgetting. In terms of the third one, Yanhui gave a detailed explanation, "It is a dropping away of my limbs and torso, a chasing off of my sensory acuity, which disperses my physical form and outs my understanding until I am the same as the Transforming Openness."[21]

Both passages above from Zhuangzi describes the journey to the Way as a shedding of the false self to unveil the true self—a protracted endeavor requiring inner effort. By inner effort, Zhuangzi doesn't stress the effort of cognitive activities of the mind. He tries to eliminate the human beings' understanding. This contrasts sharply with

[18] Confucius, *The Analects*, trans. D. C. Lau (Penguin Books, 1979), p. 136.

[19] Ibid., p. 63.

[20] Zhuangzi, *Zhuangzi: The Essential Writings,* trans. Brook Ziporyn (Hackett Publishing, Indianapolis, 2009), p. 44.

[21] Ibid., p. 49.

Emerson's view, outlined earlier, where God's power descends from above to influence people's cognition. For Emerson, grasping nature leads to an understanding of God. When the mind opens to faith, God manifests within it instantly. This process is not a prolonged struggle but a swift revelation—virtue emerges not through conflict, but as wisdom gained when the will relinquishes the self and aligns with the divine. In Daoism, however, the labor of the soul unfolds over time, distinct from Emerson's spontaneous, top-down infusion of grace.

References

1. Ralph Waldo Emerson, *The Best Read Naturalist*, ed. Michael P. Brunch and Clinton Mohs (University of Virginia Press, 2017).
2. Confucius, *The Analects*. trans. D. C. Lau (Penguin Books, 1979).
3. Ralph Waldo Emerson, *Essays and Lectures*, ed. Joel Porte (New York: Literary Classics of the United States, 1983).
4. Ralph Waldo Emerson, *The Portable Emerson*, ed. Jeffrey S. Cramer (Penguin Books, 2014).
5. Mencius, *Mengzi: With Selections from Traditional Commentaries*, trans. Bryan W. Van Norden (Hackett Publishing, Indianapolis, 2008).
6. Zhuangzi, *The Complete Works of Chuang Tzu*, trans. Burton Watson (Columbia University Press, 1968).
7. Zhuangzi, *Zhuangzi: The Essential Writings: With Selections from Traditional Commentaries*, trans. Brook Ziporyn (Hackett Publishing Company, 2009).

Chapter 12
Spiritual Practices for Self-Transcendence

Emerson and Chinese philosophy both stress spiritual practices for self-transcendence. However, there is a divergence in their understanding of spiritual practices. The former stresses outward transcendence, while the latter pays attention to inward transcendence. For Emerson, we can achieve unity with God through knowing nature. In Confucianism and Daoism, we can better know Heaven and Earth and get united with it through understanding deeply about human nature. For example, Mencius asserts, "The ten thousand things are all brought to completion by us."[1] For Mencius, we can turn inward to know Heaven and Earth instead of seeking the knowledge of nature. Emerson's stress of spiritual practices includes solitude, reading, bodily practices and expression, which he thinks can bridge humanity and God, while Chinese philosophers emphasize rituals, fasting of heart and mind, and forgetfulness, which they think can better cultivate the inner self for achieving unity with Heaven and Earth. In the following, I will make a comparative study between Emerson and Chinese philosophy in their different types of spiritual practices which include solitude, reading, bodily practices, and expression. What I would like to stress in the following is that these concepts are interpreted differently in their work. For Emerson, spiritual practices can help humanity achieve unity with God. By contrast, Confucianism and Daoism stress the unity with Dao through these practices.

1 Solitude

Emerson and Chinese philosophy both stress solitude although they have complete different interpretation of this practice. For Emerson, solitude can help humanity to be connected to God. He writes, "The simple habit of sitting alone occasionally to explore what facts of moments lie in the memory may have the effect in some more

[1] Mencius, *Mencius*, trans. Bryan W. Van Norden (Indianapolis: Hackett Publishing, 2008), p. 172.

S. Gao, *Comparative Studies Between Emerson and Chinese Philosophy*, Nature Aesthetics: Interdisciplinary and Intercultural Dialogue,
https://doi.org/10.1007/978-981-95-0421-3_12

favored hour to open to the student the kingdom of spiritual nature. He may become aware there around him roll new at this moment and inexhaustible the waters of Life; that the world he has lived in so heedless, so gross, is illumined with meaning, that every fact is magical; every atom alive, and he is the heir of it all."[2] From this quote, we can see that solitude for Emerson can help us pay attention to the vitality of life around us, which has meaning. The meaning of all lives is derived from God. Solitude for him can help us connect with God through God's revelations. He writes, "We require such a solitude as shall hold us to its revelations when we are in the street and in palaces; for most men are cowed in society, and say good things to you in private, but will not stand to them in public. But let us not be the victims of words."[3] The revelations are originated from God. For Emerson, solitude can help us receive the revelations effectively and stick to them when they are in public places. For him, we should keep a balance between solitude and society. We cannot lose sympathy towards human beings through practicing solitude. He writes, "Solitude is impracticable, and society fatal. We must keep our head in the one and our hands in the other. The conditions are met, if we keep our independence, yet do not lose our sympathy." [4]

Emerson stresses that solitude is best conducted through immersion in nature. The reason can be found in the following quote, "Nature speaks to the imagination; first, through her grand style—the hint of immense force and unity which her works convey; secondly, because her visible productions and changes are the nouns of language, and our only means of uttering the invisible thought."[5] For Emerson, nature can guide people to know God through activating people's understanding and imagination. For him, there is a direct correspondence between nature and the human mind. When mind is activated through nature, we can gain a deeper understanding of God. Confucianism also stresses solitude. However, in solitude, the activity of heart-mind still center on moral relationship with human beings. For example, the well-known Zengzi's self-reflection in solitude in the *Analects* shows that his self-reflection in solitude still center on the human beings' moral relationship among people rather than the relationship between human beings and nature. Daoism takes a different approach in solitude and it stress forgetting the moral relationship among human beings and establish the connection with Dao.

---

[2] *The Early Lectures of Ralph Waldo Emerson*, Vol. 2, ed. Stephen E. Whicher, Robert E. Spiller, Wallace E. Williams (The Belknap Press of Harvard University Press, 1964), p. 261.

[3] Ralph Waldo Emerson, *The Portable Emerson*, ed. Jeffrey S. Cramer (Penguin Books, 2014), p. 568.

[4] Ibid.

[5] Ralph Waldo Emerson, *The Best Read Naturalist*, Michael P. Branch and Clinton Mohs (University of Virgina Press, 2017), p. 205.

2 Reading

Emerson and Confucianism both stress reading while Daoism focuses on emptying the mind. For Emerson, reading refines the spirit, though Emerson cautions against scholarly seclusion. He writes, "One must be an inventor to read well. As the proverb says, 'He that would bring home the wealth of the Indies, must carry out the wealth of the Indies.' There is then creative reading as well as creative writing. When the mind is braced by labor and invention, the page of whatever book we read becomes luminous with manifold allusion."[6] Books, when guided by the reader's will, dispel chaos and usher in clarity, serving as a beacon for the soul's ascent to God. Confucianism also stresses reading. However, Confucianism focuses on reading with reverence rather reading guided by reader's will. Zhuxi, for example, once gave a detailed account of reading. Professor Guoxiang Peng has done a deep research on this aspect. He quoted Zhuxi's account of reading, "When reading people should empty their mind and pacify their emotion. Even though the meanings are proper and clear, they should not establish their own understanding first and force their own idea to make up some meaning so that they cannot see the original meaning of sages and men of virtue."[7] Zhuxi didn't advocate the creative reading as Emerson stressed. For him, reading is the way for us to get united with moral sages' words and immersed ourself in their moral feelings and world. This is a way for us to be deeply connected to Dao. By contrast, Daoism stresses emptying the mind. For them, the sages' words cannot effectively connect us with the Dao. We can grasp the Dao with empty mind rather than with understanding of the words of sages.

3 Bodily Practices

In terms of spiritual practices, both Emerson and Chinese philosophy emphasize the significant role of the body practices in cultivating virtue. Emerson highlights various forms of bodily practices as essential paths to cultivation of virtue. These include physical labor (such as farming), hiking in the woods, and direct encounter with the natural world. He discusses this in the essay of Country Life. "The importance to the intellect of exposing the body and brain to the fine mineral and imponderable agents of the air makes the chief interest in the subject. ...Walking has the best value as gymnastics for the mind. …Few men know how to take a walk. The qualifications of a professor are endurance, plain clothes, old shoes, an eye for nature, good humor, vast curiosity, good speech, good silence, and nothing too much." [8] Emerson also

[6] Ralph Waldo Emerson, *The Portable Emerson*, ed. Jeffrey S. Cramer (Penguin Books, 2014), p. 62.

[7] Guoxiang Peng, *The Confucian Tradition: Between Religion and Humanism* (Bridge 211, Alhambra, CA, 2003), p. 103.

[8] Ralph Waldo Emerson, *The Best Read Naturalist: Nature Writings of Ralph Waldo Emerson*, ed. Michael P. Branch and Clinton Mohs (University of Virgina Press, 2017), pp. 189-190.

stresses physical labor on the land and thinks that it can unveil nature's secrets, balancing its practical utility with its aesthetic depth, and serves as a crucible where mental faculties sharpen through physical struggle. In Confucianism, bodily practices encompass everyday disciplines such as proper clothing, refined demeanor, posture, and eating rituals, which help cultivate their virtue through outer form. The Confucianism' classics *The Book of Rites (Liji)* provides highly detailed prescriptions for bodily practices across all aspects of life. The disciplined body is a manifestation of moral order and harmony with the Dao. In Daoism, bodily practices are characterized by meditative exercises that unite body, mind and *qi*.

4 Expression

In spiritual practices, Emerson stresses both ways of expression: spoken language and writing. In contrast, Confucianism stresses moral action, while Daoism stresses empty mind without the medium of language. For Emerson, truth needs to be expressed in materials of spoken languages. It is not only beautiful for us to express truth by using precise language but also it is the duty of scholars of spiritual seeing and reporting. William James in his Address at the Emerson Centenary in Concord writes, "This was first half of Emerson, but only half, for genius, as he said, is insatiate for expression, and truth has to be clad in the right verbal garment. ...He was an artist whose medium was verbal and who wrought in spiritual material. This duty of spiritual seeing and reporting determined the whole tenor of his life."[9] For Emerson, language can help us get close to truth. He considers cultivation of this ability as a moral duty. In contrast, Confucianism focuses on moral action. "Mencius said, 'A great man need not keep his word nor does he necessarily see his action through to the end. He aims only at what is right."[10] "I have an insight into words. I am good at cultivating my 'flood-like ch'i."[11] In Confucianism, moral action is more important than expression by words. Zhuangzi in Daoism doesn't stress Confucian sense of moral action. He focuses more on internal action of the mind. By using his words, it is mind-fasting. In Zhuangzi, there is a dialogue between Confucius and Yan Hui in terms of mind-fasting: "Yan Hui said, 'What is the fasting of the mind?' Confucius said, 'If you merge all your intentions into a singularity, you will come to hear with the mind rather than with the ears. Further, you will come to hear with the vital energy rather than with the mind. For the ears are halted at what they hear. The mind is halted at whatever verifies its preconceptions. But the vital energy is an emptiness, a waiting for the presence of beings. The Course alone is what gathers

[9] William James, Address at the Emerson Centenary in Concord in *Memories and Studies* (Longmans, Green & Co, 1911), pp. 21-22.

[10] Mencius, *Mencius*, trans. D. C. Lau (Penguin Books, 2004), p. 90.

[11] Ibid., p. 33.

in this emptiness. And it is the emptiness that is the fasting of the mind."[12] For Zhuangzi, we can not know Dao through languages. We must cultivate the empty mind consciously so that we can be closer to Dao.

References

1. Mencius, *Mencius*, trans. Bryan W. Van Norden (Indianapolis: Hackett Publishing, 2008).
2. Ralph Waldo Emerson, *The Collected Works of Ralph Raldo Emerson*, Volume I: Nature, Addresses, and Lectures, ed. Robert E. Spiller and Alfred R. Ferguson (Cambridge, MA: Harvard University Press, 1971).
3. Ralph Waldo Emerson, *Essays and Lectures*, ed. Joel Porte (Literary Classics of the United States, 1983).
4. Guoxiang Peng, *The Confucian Tradition: Between Religion and Humanism* (Bridge 211, Alhambra, CA, 2003).
5. Ralph Waldo Emerson, *The Best Read Naturalist: Nature Writings of Ralph Waldo Emerson*, ed. Michael P. Branch & Clinton Mohs (University of Virginia Press, 2017).
6. William James, *Memories and Studies* (Longmans, Green & Co. 1911).
7. Zhuangzi, *Zhuangzi: The Essential Writings With Selections from Traditional Commentaries* (Indianapolis: Hackett Publishing Company, 2009).
8. *The Early Lectures of Ralph Waldo Emerson*, Vol. 2, ed. Stephen E. Whicher, Robert E. Spiller, Wallace E. Williams (The Belknap Press of Harvard University Press, 1964).

[12] Zhuangzi, *Zhuangzi:The Essential Writings With Selections from Traditional Commentariesi* (Indianapolis: Hackett Publishing Company, 2009), p. 26.

Chapter 13
Trinity, Unity, and Aesthetic Feelings

In Emerson and Chinese Philosophy, there exists a highest state of virtue, which is to live in harmony with nature. It is also the ideal state of self-transcendence. In Emerson, the highest state of virtue is to form a trinity among nature, humanity, and God. In Chinese philosophy, it is to achieve unity between human beings and nature and the expression for this unity in Chinese is 天人合一(*tian ren he yi).* When attaining self-transcendence, people will experience the aesthetic feelings. For Emerson, he uses the term of ecstasy (sometimes sublime) to express this feeling. Chinese philosophy uses the term of joy to express it. For Emerson, aesthetic feeling of ecstasy is connected to ultimate values of truth, goodness, and beauty. The feeling of joy in Confucianism is directly from human beings' virtue. Daoism also stresses joy but in the both positive and negative way. For Daoism, the supreme joy is "no joy", which is positive feeling through being negative. This feeling is also connected to virtue and the highest state of virtue is the same as Confucianism: living in harmony with nature. Emerson's mysticism, as noted by Cornel West, shines through silence and celebration, "Most of Emerson's mysticism rests upon his silent yet discernible sense of being jubilant and celebratory that he is alive. He discloses a sense of being contented and full of joy that he 'dwells' in the house of beings."[1] Unlike Zhuangzi, who finds mystery in free wandering within an imagined "land of nothingness" beyond time and reality, Emerson sees nature not as an ultimate material being but as spirit, its essence resolving into thought. Confucian mysticism emerges from the fusion of moral vitality with the energy of Heaven and Earth, sharing with Emerson a reverence for natural space as a conduit for this experience. This chapter will give a philosophical interpretation of the concept of trinity, unity and aesthetic feelings of ecstasy and joy in Emerson and Chinese philosophy.

---

[1] Cornel West, *The American Evasion of Philosophy: A Genealogy of Pragmatism* (Madison: University of Wisconsin Press, 1989), p. 25.

S. Gao, *Comparative Studies Between Emerson and Chinese Philosophy*, Nature Aesthetics: Interdisciplinary and Intercultural Dialogue,
https://doi.org/10.1007/978-981-95-0421-3_13

1 The New Trinity and the Unity of Heaven and Humanity

Emerson reconstructs the traditional trinity in Christianity with a new trinity, which is the unity among humanity, nature, and God. The essence of unity among humanity, nature, and God is human beings' attainment of self-transcendence towards the ultimate values. The emotion of sublime arises when humanity merges with the ultimate values of truth, goodness, and beauty. He writes in his essay "Oversoul", "We distinguish the announcements of the soul, its manifestations of its own nature, by the term *Revelation.* These are always attended by the emotion of the sublime. For this communication is an influx of the Divine mind into our mind. It is an ebb of the individual rivulet before the flowing surges of the sea of life."[2] For Emerson, God is in essence the consciousness of the ultimate values and it has the power to inspire people to know these values. When human beings achieve self-transcendence through inspiration, the aesthetic emotion of sublime will be generated.

In this unity, Emerson dissolves secular ties, perceiving all existence as spirit. His often quoted passage of "transparent eyeball" best illustrates this, "Standing on the bare ground,—my head bathed by the blithe air and uplifted into infinite space,—all mean egotism vanishes. I become a transparent eyeball; I am nothing; I see all; the currents of the Universal Being circulate through me; I am part or parcel of God."[3] Alone in nature, Emerson sheds social bonds, and the material world ascends to spirit, flowing as divine energy. For Emerson, the relationship between human beings and God is most important and nature serves as the medium for people to know God. Human relationship is governed by the relationship with God.

Chinese philosophy, by contrast, centers on the unity between Heaven and human Beings. Nature is not the medium for people to know Heaven. Instead, we can better know Heaven through interrelationship with human beings. Heaven and Earth in Chinese philosophy is deeply connected to the world of human beings. Song Confucian Zhangzai writes, "Heaven is my father and Earth is my mother, and even such a small creature as I finds an intimate place in their midst."[4] Here, human beings and nature form a familial bond under Heaven and Earth. Zhuangzi echoes this in his vision of supreme virtue: "Thus, in the age of perfect Virtuosity their actions were solid and full but their gaze was distant and blank. For in those days, there were no paths or trails through the mountains, no boats or bridges over the ponds; all creatures lived together, merging their territories into one another. "[5] This unity roots itself in a lived space, not a spiritual abstraction as in Emerson, harmonizing humanity with heaven through tangible existence.

---

[2] Ralph Waldo Emerson, *The Portable Emerson*, ed. Jeffrey S. Cramer (Penguin Books, 2014), p. 217.

[3] Ibid., p. 20.

[4] *A Source Book in Chinese Philosophy*, trans. Wing-tsit Chan (Princeton University Press, 1963), p. 497.

[5] Zhuangzi, *Zhuangzi: The Essential Writings with Selections from Traditional Commentaries,* trans. Brook Ziporyn, (Indianapolis: Hackett Publishing, 2009), p. 61.

2 Ecstasy and Joy

Ecstasy and joy are the aesthetic feelings which are stressed by Emerson and Chinese philosophy when people achieve self-transcendence. In this section, I will make a comparative study between Emerson and Chinese philosophy in their understanding of aesthetic feeling of joy from the relationship between intensity of feelings and the ultimate values.

In terms of the intensity of the feeling of joy, Emerson's concept of ecstasy is intense but transient, while the feeling of joy in Confucianism and Daoism is peaceful and enduring. Emerson's ecstasy erupts momentarily, dissolving the self into God's spirit. Andrew. M. Barry, citing William James, notes its traits—ineffability, poetic quality, and transience: "Emerson loses his sense of self and feels alienated from others and his surroundings. He experiences a sense of elevation and of being pervaded by the 'currents of the Universal Being'".[6] Arthur Versluis terms this "immediatism," a spontaneous awakening requiring no prolonged effort, "Immediatism is not the same as perennialism, universalism, or traditionalism—it refers to a perspective emphasizing direct and immediate access to spiritual insight, and does not carry along with it ancillary doctrines that valorize ancient religions and cultures or for that matter, contemporary world religious traditions."[7] Emerson underscores the relationship between ecstasy and the ultimate values of Truth, Goodness, and Beauty. This relationship is best demonstrated in his essay "The Transcendentalist". He writes, "The Transcendentalist adopts the whole connection of spiritual doctrine. He believes in miracle, in the perpetual openness of the human mind to new influx of light and power; he believes in inspiration, and in ecstasy....But this class are not sufficiently characterized, if we omit to add that they are lovers and worshippers of Beauty. In the eternal trinity of Truth, Goodness, and Beauty each in its perfection including the three, they prefer to make Beauty the sign and head."[8] This passage reveals that Emerson's concept of ecstasy is intimately tied to the pursuit of ultimate values. For him, ecstasy is not merely an emotional state of rapture, but a profound spiritual experience in which the individual mind opens itself to ultimate values and power.

Confucian joy arises from moral harmony with Heaven and Earth. In *The Analects*, there is a dialogue between Ssu-ma Niu and Confucius about the gentleman. The Master said, "The gentleman is free from worries and fears."[9] This calm joy—freedom from worry—contrasts with Emerson's fervor. The feeling of joy is also mentioned in the following passage, "The Master said, 'How admirable Hui is! Living in a mean dwelling on a bowlful of rice and a ladleful of water is a hardship

[6] Andrews M. Barry, "That Which Was Ecstasy Shall Become Daily Bread," *Religions*, 2017, 8(4), 75.

[7] Arthur Versluis, *American Gurus: From Transcendentalism to New Age Religion* (Oxford University Press, 2014), p.4.

[8] Ralph Waldo Emerson, *The Portable Emerson*, ed. Jeffrey S. Cramer (Penguin Books, 2014), pp. 115-125.

[9] Confucius, *The Analects*, trans. D. C. Lau (Penguin Books, 1979), p. 113.

most men would find intolerable, but Hui does not allow this to affect his joy. How admirable Hui is!"[10] From this passage, we can see that joy in confucianism is the outcome of practicing virtue.

Zhuangzi's "heavenly joy" shuns purposeful pursuits of Confucian sense of virtue, "I take inaction to be true happiness, but ordinary people think it is a bitter thing. I say: perfect happiness knows no happiness, perfect praise knows no praise. The world can't decide what is right and what is wrong. And yet inaction can decide this. Perfect happiness, keeping alive—only inaction gets you close to this!"[11] He writes, "So it is said, for him who understands Heavenly Joy, life is the working of Heaven; death is the transformation of things. In stillness, he and the yin share a single Virtue; in motion, he and the yang share a single flow. Thus he who understands Heavenly joy incurs no wrath from Heaven, no opposition from man, no entanglement from things, no blame from the spirits."[12] This tranquil joy, born of oneness with the Dao, eschews moral or sensory striving, differing from Emerson's ecstatic uplift.

Aesthetic feelings of ecstasy and joy are deeply connected to ultimate values. For Emerson, scientific truth is the important value for aesthetic feelings. Science can help us better understand nature and ourselves. He states, "The motive of science was the extention of man, on all sides, into Nature, till his hands should touch the stars, his eyes see through the earth, his ears understand the language of beast and bird, and the sense of the wind; and through his sympathy, heaven and earth should talk with him." [13]For Emerson, the aesthetic feelings are directly derived from knowing the scientific truth. However, for Confucianism and Daoism, joy is mostly related to knowing the truth in the society of human beings.

References

1. *A Source Book in Chinese Philosophy*, trans. Wing-tsit Chan (Princeton University Press, 1963).
2. Andrews M. Barry, "That Which Was Ecstasy Shall Become Daily Bread", *Religions*, 2017, 8 (4).
3. Confucius, *The Analects*, trans. D. C. Lau (Penguin Books, 1979).
4. Cornel West, *The American Evasion of Philosophy: A Genealogy of Pragmatism* (Madison: University of Wisconsin Press, 1989).
5. Ralph Waldo Emerson, "The Methods of Nature" in *The Collected Works of Ralph Waldo Emerson*, Volume I: Nature, Addresses, and Lectures, ed. Robert E. Spiller and Alfred R. Ferguson (Cambirdge, MA: Harvard University Press, 1971).
6. Haydan Washington, *A Sense of Wonder Towards Nature: Healing the Planet through Belonging* (Routledge, 2018).
7. John Nolt, "Hope, Self-transcendence and Environmental Ethics", *Inquiry* 53, no. 2 (2010).

[10] Ibid,. p. 82.

[11] *The Complete Works of Chuang Tzu*, trans. Burton Watson (New York: Columbia University press, 1968), p. 191.

[12] Ibid., p. 144.

[13] Ralph Waldo Emerson, *Essays and Lectures*, ed. Joel Portel (New York: Literary Classics of the United States, 1983), p. 1100.

8. Martha C. Nussbaum, *Upheavals of Thought: The Intelligence of Emotions* (Cambridge University Press, 2001).
9. Ralph Waldo Emerson, *Essays and Lectures*, ed. Joel Porte (Literary Classics of the United States, 1983).
10. Ralph Waldo Emerson, *The Portable Emerson*, ed. Jeffrey S. Cramer (Penguin Books, 2014).
11. Shan Gao, "Can Chinese Philosophy Embrace Wilderness?", *Environmental Ethics* 39, no. 2 (Summer 2017).
12. Zhuangzi, *The Complete Works of Chuang Tzu* (Columbia University Press, 1968).
13. Arthur Versluis, *American Gurus: From Transcendentalism to New Age Religion* (Oxford: Oxford University Press, 2014).
14. Zhuangzi, *Zhuangzi: The Essential Writings with Selections from Traditional Commentaries*, trans. Brook Ziporyn (Hackett Publishing Company, 2009).

Part IV
Place

In this part, I will focus on the concept of place. Emerson and Chinese philosophers's metaphysical interpretation of nature is value-laden. That is the one of the important reasons why they both pursue spirituality through living in harmony with nature. Their spirituality has great impact on their experience of place. Emerson's metaphysics of nature brings a new conception of wilderness. Wilderness is no longer regarded as meaningless space rather it is considered as place with values. The metaphysics in Chinese philosophy generate topophilia, which refers to attachment to cultivated places such as farm, gardens and culturalized mountains.

Chapter 14
Place

In the last chapter of this book, I will focus on the concept of place, which has been neglected by many Western philosophers due to their metaphysical interpretation of the concept of nature. In the past decades, the concept of "place" has been a focal point across disciplines such as philosophy, religion, literature, and geography. Notable thinkers, including Chinese geographer Yi-Fu Tuan, Jeff Malpas, and Paul B. Thompson, have deepened this discourse by highlighting the moral, aesthetic, and existential dimensions of human engagement with specific places. The concept of place includes four elements: (1) Geological space and material things in the space. Place is fundamentally linked to physical space. This physical space is either material or virtual as AI has created virtual space based on material nature. However it is not the abstract, geometric framework outlined by René Descartes. Rather than it is a space that has value. (2) Activities of human beings and non-human beings in the space. Place is deeply related to the activities and events of human beings and non-human beings. (3) Body. The concept of place is inseparable from body of human beings and non-human beings. (4) Aesthetic and ethical dimensions. Place is deeply connected to aesthetic appreciation of human beings and their pursuit of virtue, and I will discuss this later. This chapter explores the concept of place from three dimensions: metaphysical interpretation of nature and place; aesthetics, ethics, and place; conservation and place.

1 Metaphysical Interpretation of Nature and Place

Metaphysical interpretation of nature has direct influence on people's sense of place. Many Western philosophers' metaphysical interpretations of nature as I have examined in part I tend to neglect the concept of place. The main reason is that their metaphysical framework doesn't embrace the concept of the place. Their metaphysical framework is characterized by two features: (1) The ultimate reality doesn't lie

S. Gao, *Comparative Studies Between Emerson and Chinese Philosophy*, Nature Aesthetics: Interdisciplinary and Intercultural Dialogue,
https://doi.org/10.1007/978-981-95-0421-3_14

in the material world we live in. Instead, the transcendental world is real; (2) Body is inferior to the mind, which can guide people to the truth, goodness, and beauty. These features will cause the result of neglecting the concept of place. I will use the example of the representative philosophers' interpretation of nature to illustrate how they neglect the concept of place.

First, in terms of the philosophers who integrate value with nature such as Aristotle, Plotinus, Augustine, and Rousseau, their interpretations of nature tend to focus on transcendental reality, which is also termed as God or the One. God belongs to the transcendental world and it has the ultimate values of the truth, goodness, and beauty. Material world for them is inferior to this transcendental world. Their ethics and aesthetics are related to appreciation of this world instead of the world we live in everyday life. For example, Aristotle thinks that the best life is a contemplative life, which is directed towards thinking and knowing the transcendental world. For Plotinus and Augustine, contemplating God or the One is metaphysically important. For Rousseau, the ultimate values lie in God. Most of them stress develop our mind to know God and they tend to ignore the importance of bodily cognition.

Second, for the philosophers who separate value from nature with regards to their metaphysical interpretation of nature such as Epicurus, Spinoza, Descartes and Kant. They don't have aesthetic appreciation of nature although they stress knowing nature. Spinoza and Descartes' metaphysical interpretation of nature stresses sciences such as mathematics, geometry, and physics. Their interpretation neglects our bodily interaction with nature and activities in nature, which is important to develop a sense of place. The values they stress are related only to human beings' welfare or good. When nature is the object of scientific observation, it is difficult for people to develop a sense of place.

Third, Nietzsche's metaphysical interpretation of nature which is based on the core concept of will to power provides the possibility for valuing nature aesthetically. However, his denial of existence of truth, goodness, and beauty in nature makes him difficult to embrace the idea of place. For him, appreciate nature aesthetically can help people overcome the pressure from human society. However, the beauty of nature cannot become his spiritual pursuit as Emerson does.

Compared with many Western philosophers, the metaphysical interpretation of nature in Emerson and Chinese philosophy provide solid foundation for embracing the concept of place. Emerson's interpretation of nature makes wilderness become a place. Nash has offered a deep insight on this. He points out "Civilization created wilderness."[1] Chinese philosophy's metaphysical interpretation of Heaven and Earth makes the agricultural land, gardens, mountains and waters become a place in Chinese culture.

[1] Roderick Nash, *Wilderness and the American Mind*, third edition (Yale University Press, 1967), xiii.

2 Aesthetics, Ethics, and Place

Aesthetic appreciation and ethical valuing the material things are important elements of place. Many Western philosophers' metaphysical interpretations of nature as I examined earlier in this book neglect the concept of place since their metaphysics doesn't support aesthetic appreciation and valuing the material things. For example, Spinoza doesn't have an aesthetic appreciation of material things. For him, extension is one of God's attributes and material things are understood by this attribute. Material things are not beautiful since God creates all things without following any end. He strongly objects that people use value terms to describe God. Descartes, like Spinoza, also doesn't have any aesthetic appreciation of material things. His explanation of material things is mathematical and geometrical. It is not related to human beings' experience of material things through activities. By aesthetics, I refer to aesthetic appreciation of nature with bodily participation. Body is an important element in people's sense of place. Without bodily participation, it is difficult for people to have a sense of place. In this aspect, many Western philosophers as I discussed above tends to ignore the role of body in their metaphysics. That is one of the deeper reasons why they neglect the concept of place. By ethics, I refer to human beings' pursuit of virtue, which has guided their aesthetic appreciation of material things. For example, Emerson's aesthetic appreciation of wilderness is deeply related to virtue as I have examined earlier. Wilderness can practice human beings cognitive faculties of mind which are important element of virtue. In Chinese philosophy, the aesthetic appreciation of farming land and gardens is also connected to the pursuit of human beings' relationships, which is the foundation for cultivating virtue. One point I would like to stress here is that people's aesthetic appreciation of place will change over time when people have developed the pursuit of new self and virtue. For example, when virtual world has becoming developed these days, people's aesthetic appreciation is enhanced in the virtual space, they may develop a new virtual self and virtue in the new world. At that time, virtual space will become a new place. David J. Chalmers in his recent book titled *Reality+Virtual worlds and the Problems of Philosophy*, expresses this possibility. He writes, "In the future, we may come up with a new technology we can call Terraform Reality. Terraform Reality allows us to turn exoplanets in the nonvirtual world into habitable environments full of beautiful places and engaging activities".[2] Space and time in virtual space are very different from the non-virtual world. Many activities and events have been conducted in this space. It provides a possible way of life and construction of new virtual self.

[2] David J. Chalmers, *Reality+ Virtual Worlds and the Problems of Philosophy* (W. W. Norton & Company, 2022), p. 311.

3 Conservation and Place

The concept of place appears frequently in today's conservation practices. It is often used in the conservation of various heritages such as agricultural cultural heritage, industrial heritage, and so on. In practice, when human beings make decisions, they have to solve the following questions: How to evaluate diverse values of different places? Which place is more valuable, wilderness, agricultural land, outer space, or industrial land? How different cultures value different landscapes? How can we find objective ground for valuing different places around the world? All of these questions need philosophical justification. Without philosophical justification, many policies will not motivate people to protect these valuable places. This book ends with the chapter of place and I aim at providing the guidance for law and policy makers in making decisions for conservation. Right now, China has established the national park law. I think it is a great breakthrough. However, a large areas of wilderness have not been well protected in China. It has not been well deffined and researched concept. Through the comparative studies, we can get more philosophical insight about the importance of wilderness, a concept which has been neglected for thousands of years. Through intercultural dialogue, we can get a new insight about the value in wilderness and its importance in defining the self. This insight will have the potential to develop a new value direction in China and wilderness becomes a meaningful place.

References

1. Roderick Nash, *Wilderness and the American Mind*, third edition (Yale University Press, 1967).
2. David J. Chalmers, *Reality+ Virtual Worlds and the Problems of Philosophy* (W. W. Norton & Company, 2022), p. 311.

Chapter 15
Emerson and Wilderness as a Place

In Emerson, wilderness is regarded as a place, which has value. Wilderness is one type of landscapes which include farming land, city, gardens and so on. In the history of Western philosophy, wilderness as material nature is usally not valued. However, Emerson's metaphysical interpretation of nature changes this perception. Wilderness is valued for its connection with the divine God. He writes, "Things admit of being used as symbols because nature is a symbol, in the whole, and in every part. Every line we can draw in the sand has expression; and there is no body without its spirit or genius."[1] For Emerson, wilderness as one type of nature is the medium for human beings to know God. God is also immanent in all of the objects of nature. It is through this relationship, wilderness is aesthetically appreciated and ethically valued as a place. However, there is one point I would like to stress here is that although Emerson promotes aesthetic appreciation of wilderness, which becomes a place for people rather than geological space, Emerson doesn't promote the conservation of these places since his appreciation of wilderness is still second to his appreciation of God. He influences many artists and philosophers, which have direct influence on creation of American wilderness act and conservation practices. In this chapter, I will first discuss Emerson's aesthetic appreciation of wilderness. Second, I will explore how Emerson influences artists and philosopher in the idea of wilderness as a meaningful place.

[1] Ralph Waldo Emerson, *The Portable Emerson*, ed. Jeffrey S. Cramer (Penguin Books, 2014), p. 242.

S. Gao, *Comparative Studies Between Emerson and Chinese Philosophy*, Nature Aesthetics: Interdisciplinary and Intercultural Dialogue,
https://doi.org/10.1007/978-981-95-0421-3_15

1 Emerson's Nature Aesthetics

Emerson views material nature as inherently beautiful, a perspective rooted in his metaphysical belief in God as the ultimate source of truth, goodness, and beauty. I will explore his natural aesthetics through three dimensions: (1) aesthetic appreciation of all kinds of nature; (2) nature as a symbol; (3) the role of science in aesthetic appreciation of nature. These aspects highlight Emerson's distinct approach, contrasting with both scientific rationality and Chinese cultural perspectives.

First, aesthetic appreciation of all kinds of nature. Emerson perceives nature's beauty holistically, a view he articulates in his writings and embodies in his life. He writes, "The Universe is a more amazing puzzle than ever as you glance along this bewildering series of animated forms, the hazy butterflies, the carved shells, the birds, beasts, fishes, insects, snakes, and the upheaving principle of life everywhere incipient in the very rock aping organized forms."[2] Unlike modern environmental aesthetician Allen Carlson, who grounds nature's beauty in scientific rationality, Emerson attributes it to God's spiritual laws, rendering every element—from the grotesque to the sublime—beautiful as a divine expression. Emerson's aesthetic appreciation includes living forms of all things, time, invisible air, various landscapes and so on. Emerson celebrates nature's temporal and spatial dynamics, as seen in his essay "Nature", "The inhabitants of cities suppose that the country landscape is pleasant only half the year. I please myself with the graces of the winter scenery, and believe that we are as much touched by it as by the genial influences of summer."[3] Emerson appreciates various life forms unfolding in different seasons. He also expresses an aesthetic appreciation of invisible air. He writes, "In good health, the air is a cordial of incredible virtue. Crossing a bare common, in snow puddles, at twilight, under a clouded sky, without having in my thoughts any occurrence of special good fortune, I have enjoyed a perfect exhilaration."[4] Emerson's aesthetic appreciation of air is informed by scientific knowledge. He often quotes some scientists' interpretation of air or other natural elements in his essays."

Emerson's appreciation also includes various landscapes such as farming land, gardens and wilderness. His appreciation of farming land can be seen in this passage "The heavens change every moment, and reflect their glory or gloom on the plains beneath. The state of the crop in the surrounding farms alters the expression of the earth from week to week. …By water-courses, the variety is greater. In July, the blue pontederia or pickerel-weed blooms in large beds in the shallow parts of our pleasant river, and swarms with yellow butterflies in continual motion. Art cannot rival this pomp of purple and gold. Indeed the river is perpetual gala, and boasts each month a new ornament."[5] These passages fuse time with the vitality of nature's forms, spanning cultivated and wild spaces.

---

[2] Robert D. Richardson Jr., *Emerson: The Mind on Fire* (University of California Press, 1995), p. 141.

[3] Emerson, *The Portable Emerson*, ed. Jeffrey S. Cramer (Penguin Books, 2014), p. 24.

[4] Ibid., pp. 19–20.

[5] Ibid., p. 24.

Second, Emerson regards natural beauty as symbolic, not ultimate, serving as a conduit to God's eternal beauty. He writes, "Truth has not single victories; all things are its organs,—not only dust and stones, but errors and lies."[6]; "Beauty, in its largest and profoundest sense, is one expression for the universe. God is the all-fair. Truth, and goodness, and beauty, are but different faces of the same All. But beauty in nature is not ultimate. It is the herald of inward and eternal beauty, and is not alone a solid and satisfactory good. It mast stand as a part, and not as yet the last or highest expression of the final cause of nature."[7] Material Nature, as God's creation, points beyond sensory pleasure to divine truth, rendering superficial pursuits incomplete without spiritual ascent. Material nature for Emerson is not the ultimate reality and it is only the production of ultimate reality, which is God. Nature serves as a symbol for us to understand God.

Third, science plays a very important role in Emerson's aesthetic appreciation of nature. We can find many examples from his essays and lectures. In his lecture on Country Life (1858), he uses physiologist Malphigi's scientific knowledge to support his aesthetic appreciation of air. "The importance to the intellect of exposing the body and brain to the fine mineral and imponderable agents of the air makes the chief interest in the subject."[8] For Emerson, science can enhance our aesthetic appreciation of nature, but he also warns that science itself will be limited if we cannot integrate it with human beings' imagination. He has a special term for this insight: joyous science. He puts forward this term in his essay "The scholar", "Beauty, the inspirer, the cheerful festal principle, the leader of gods and men, which draws by being beautiful, and not by considerations of advantage, comes in and puts a new face on the world. I think the peculiar office of scholars in a careful and gloomy generation is to be (as the poets were called in the Middle Ages). Professors of the Joyous Science, detectors and delineators of occult symmetries and unpublished beauties; heralds of civility, nobility, learning and wisdom; affirmers of the one law, yet as those who should affirm it in music and dancing; expressors themselves of that firm and cheerful temper, infinitely removed from sadness, which reigns through the kingdoms of chemistry, vegetation, and animal life."[9] For Emerson, the beauty of nature itself has a power to awaken people's imagination and reason to know the final cause of nature. Science itself must work together with this imagination so that we can know the final cause of nature.

---

6 Ibid., p. 203.

7 Ibid., p. 27.

8 Emerson, *The Best Read Naturalist*, ed. Michael P. Branch and Clinton Mohs (University of Virginia Press, 2017), p. 189.

9 Emerson, *Lectures and Biographical Sketches* (Cambridge, Mass.: The Riverside Press, 1895), pp. 250–251.

2 Emerson's Influence on Wilderness as a Place

Although Emerson's aesthetic appreciation of nature includes different kinds of landscapes, he regards wilderness as being most beautiful landscapes. His aesthetic appreciation of wilderness is deeply connected to his love towards them. There is a metaphysical reason for his love towards wilderness which is best expressed in the following passage. He writes, "There is deep reason for the love of nature that has characterized the highest minds. The soul and the body of things are harmonized; therefore the deeper is a man's insight into the spiritual laws the more intense will be his love of the works of nature."[10] This contrasts with the prominent Chinese-American humanistic geographer Yi-Fu Duan's idea of topophilia through everyday life experiences, as Emerson's affinity stems from the soul's connection to divine laws. For Emerson, even destructive forces evoke love. His respect for these laws is demonstrated in his essay titled Prudence. He writes, "Prudence does not go behind nature, and ask whence it is. It takes the laws of the world, whereby man's being is conditioned, as they are, and keeps these laws, that it may enjoy their proper good. It respects space and time, climate, want, sleep, the law of polarity, growth, and death."[11] For Emerson, there is authority in nature. This authority lies in its moral power which is intimately connected with scientific facts. He will not focus on some particular things in nature and give a deep research. Instead, he will respect the spiritual law in nature. Emerson's nature aesthetics profoundly inspires the American artists and nature writers of his time, while also shaping the nation's wilderness conservation movement. Ultimately, literature, art, and conservation forges America's sense of place, transforming the wilderness from a meaningless geographical space into a place imbued with values and meaning.

2.1 Emerson's Influence on Nineteenth-Century Wilderness Art in America

Emerson's ideas directly inspired nineteenth-century American landscape painting, notably the Hudson River School and Luminism. He writes, Professor Guardiano remarks that "It is a noteworthy historical fact that a clear intention of this first original tradition of painting in the United States was precisely to capture and present the American wilderness in its wealth and diversity."[12] To depict this beauty, painters became hikers and mountaineers, immersing themselves in nature. This appreciation was not inherent in early American culture. Environmental historian Roderick

[10] Emerson, *The Best Read Naturalist,* ed. Michael P. Branch and Clinton Mohs (University of Virginia Press, 2017), p. 61.

[11] Ralph Waldo Emerson, *Essays and Lectures,* ed. Joel Porte (New York: Literary Classics of the United States, 1983), p. 539.

[12] Nicholas L. Guardiano, *Aesthetic Transcendentalism in Emerson, Peirce, and Nineteenth-Century American Landscape Painting* (Lexington Books, 2017), p. 21.

Nash explains in his book titled *Wilderness and the American Mind* that European settlers initially feared the wilderness due to religious and frontier influences. He cites Michael Wigglesworth's view on wilderness, "a waste and howling wilderness / Where non inhabited / But hellish fiends, and brutish men / That Devils worshiped."[13] Nash writes, "Insofar as the westward expansion of civilization was thought good, wilderness was bad. It was construed as much a barrier to progress, prosperity, and power as it was to godliness"[14] Early Americans often framed their relationship with wilderness as a military conquest, a stark contrast to later transcendentalist reverence. The Hudson River School transformed this perception, reflecting Emerson's core beliefs: beauty as the universe's highest principle and nature's ceaseless creativity. Nicholas L. Guardiano, in his book titled *Aesthetic Transcendentalism in Emerson, Peirce, and Nineteenth-Century American Landscape painting*[15], illustrates this through specific artworks. In this book, he points out that in Emerson's metaphysics, beauty is regarded as the highest metaphysical principle: Martin Johnson Heade's Thunderstorm over Narragansett Bay exemplifies this. Dark clouds, lightning, seagulls, and a tranquil human figure amid an impending storm convey wilderness as a divine, beautiful creation, not a threat. Unlike urban storms, wilderness tempests evoke contemplation, aligning with Emerson's view that "beauty is the creator of the universe,"[16] connecting humanity to God without fear. This contrasts with Kant's sublime, where nature's forces provoke awe through rational triumph over sensory fear, not inherent beauty.[17] This beauty stems from nature's diverse, every-changing forms, such as mountains, canyons, lakes, trees, and storms, which are depicted with high realism. Guardiano also mentions Asher Brown Durand's Learning from Nature, he points out that The Strath of Vermont captures a fallen tree's details—curves, bark, and dead leaves in varied browns—within a broader scene of sky and mountains, reflecting nature's subtle unity. This is the reflection of influence from Emerson's metaphysics. For him, the efficient nature always produces forms, which are all beautiful. There is nothing ugly or false in nature.

---

[13] Roderick Nash, *Wilderness and the American Mind*, third edition (Yale University press, 1967), p. 36.

[14] Ibid., p. 40.

[15] Nicholas L. Guardiano, *Aesthetic Transcendentalism in Emerson, Peirce, and Nineteenth-Century American Landscape Painting* (Lexington Books, 2017), pp. 1–37.

[16] Ibid., p. 14.

[17] Immanuel Kant, *Critique of Judgement*, trans. Werner S. Pluhar (Hackett Publishing, 1790/1987), pp. 120–122.

2.2 *Emerson's Influence on Wilderness Literature in America*

Emerson has strong influence on wilderness literature in America. This literature plays more important role in perception of wilderness as a place. Emerson's transcendentalism influenced Henry David Thoreau and John Muir, shaping wilderness literature. In his 1871 diary, Emerson lists Muir among "my team," alongside Thoreau and others.[18] While Thoreau, as Emerson's student, directly absorbed his ideas, Muir's transcendentalism predated their 1871 meeting in the Nevada Mountains, where they explored the Mariposa Redwood Forest together.[19] Emerson's influences on Thoreau and John Muir are reflected in two aspects.

First, Science and Wilderness Aesthetics: Thoreau and Muir pursued wilderness truth through science, blending it with beauty and goodness. Muir, rejecting his father's biblical fundamentalism, embraced natural theology, studying botany, geology, and ecology to enhance his appreciation. Donald Worster summarizes Muir's view at ages 35–40: "Yet Muir's rejection of Christian fundamentalism did not imply atheism or even doubt about the moral ideals he had been raised on. ...Every mountain seemed to testify to the ethical principles taught in the Sermon on the Mount. A few more years of mountain science would diminish such admiring references to Jesus or the Christian religion, but he never would renounce a belief in God (or some god-like force flowing through the natural world, flowing through the ice, and creating a world fit for life)."[20] Muir saw mountains, avalanches, and gales as beautiful manifestations of divine creativity, fearlessly hiking Alaska's glaciers, which he viewed as life-generating force, "'What we in our faithless ignorance and fear call destruction,' he wrote, 'is creation finer and finer.'"[21] This echoes Emerson's belief in nature's perpetual creation.

Second, wilderness appreciation and meaning in life: Emerson's nature aesthetics has profound influence on Thoreau. He regards wilderness appreciation as vital to the pursuit of meaning in life. In Walden, he writes, "I went to the woods because I wished to live deliberately, to front only the essential facts of life, and see if I could not learn what it had to teach, and not, when came to die, discover that I had not lived".[22] Thoreau provides a very detailed description of wilderness including sound, air, plants, animals, and so on. His detailed observation is a reflection of his pursuit of meaning in life in wilderness appreciation. Thoreau brings Emerson's nature aesthetics into practice. Through his meticulous observation and appreciation of the biodiversity of life in nature, especially in the wilderness, he fulfills his pursuit of the meaning of life. The works he completed in the form of literary narrative has

[18] Emerson, *The Journals and Miscellaneous Notebooks of Ralph Waldo Emerson*, vol. 16. Ed. Ronald A. Bosco and Glen M. Johnson (Harvard University Press, 1982), p. 235.

[19] Donald Worster, *A Passion for Nature: The Life of John Muir* (Oxford University Press, 2008), p. 202.

[20] Ibid., p. 202.

[21] Ibid., p. 272.

[22] Henry David Thoreau, *Walden or Life in the Woods and Civil Disobedience* (Signet Classics, 1999), p. 74.

profoundly shaped a sense of place, which directly influences the protection of the wilderness surrounding Walden Pond and also shapes the American public's aesthetic attitudes toward wilderness. However, it must be emphasized that the original inspiration for Thoreau's pursuit of the meaning of life achieved in the wilderness came from Emerson's reinterpretation of nature, particularly the wilderness, and that this reinterpretation gives rise to wilderness aesthetics.

2.3 Emerson's Influence on Wilderness Conservation

Emerson has indirect influence on wilderness conservation. Among three transcendentalists of Emerson, Thoreau, and Muir, Muir has direct influence on wilderness conservation through his activism. His Sierra Club leadership from 1892 aimed to "exploring, enjoying and rendering accessible the mountain regions of the Pacific Coast",[23] embodying transcendentalist ideals. Emerson's transcendental vision greatly promotes the practice of Thoreau and Muir. Their practice has direct influence on enact of 1964 American Wilderness Act, which protects "undeveloped Federal land retaining its primeval character and influence, without permanent improvements or human habitation, which is protected and managed so as to preserve its natural conditions."[24] Nash notes its arduous passage—nine hearings from 1957 to 1964, 6000 pages of testimony, and 66 resubmissions—with Thoreau and Muir frequently cited.[25]

Although Emerson influences the perception of wilderness as a place, his idea of home prevents him from taking action to wilderness conservation. Emerson's focus on human self-improvement through nature's divine symbolism contrasts with modern environmental philosophers like Holmes Rolston III, and Aldo Leopold, who see wilderness as intrinsically valuable. Rolston, in *Philosophy Gone Wild*, argues that wildness is a place.: "Wildness is a Place of *Roots* in Historical and Ongoing Senses;......Wildness Is Place of *Neighbors* in a Sense Gone Wild; ...Wildness Is a Place Where We Encounter *Aliens*."[26] For Rolston, wildness is not just a meaningless space rather it is a place with meaning. He has strong aesthetic appreciation towards it. Leopold, in his book titled *A Sand County Almanac*, expresses his aesthetic appreciation of wilderness in his land ethics, "It is inconceivable to me that an ethical obligation to land can exist without love, respect, and admiration for land, and a high regard for its value."[27] For Leopold, land as a ecological community has value for

---

[23] Roderick Nash, *Wilderness and the American Mind*, Third edition (Yale University Press, 2014), p. 132.

[24] Wilderness Act, Public Law 88-577, 88th Congress, 1964, Sec. 2 (c).

[25] John Copeland Nagle, "The Spiritual Values of Wilderness, "35 Envtl. L. 955 (2005).

[26] Holmes Rolston III, *Philosophy Gone Wild* (Prometheus Books, 1989), pp. 122–126.

[27] Aldo Leopold, *A Sand County Almanac* (Oxford University Press, 1949), p. 223.

its own sake. We have ethical obligation to conservation of this value. This obligation is based on aesthetic appreciation of wilderness. Compared with Emerson, Aldo Leopold regards the land itself as having philosophical value.

References

1. Ralph Waldo Emerson, *The Portable Emerson*, ed. Jeffrey S. Cramer (Penguin Books, 2014).
2. Robert D. Ricardson Jr., *Emerson: The Mind on Fire* (University of California Press, 1995).
3. Ralph Waldo Emerson, *The Best Read Naturalist*, ed. Michael P. Branch & Clinton Mohs (University of Virginia Press, 2017).
4. Emerson, *Lectures and Biographical Sketches* (Cambridge, Mass.: The Riverside Press, 1895).
5. Ralph Waldo Emerson, *Essays and Lectures*, ed. Joel Porte, (New York: Literary Classics of the United States, 1983).
6. Emerson, *The Journals and Miscellaneous Notebooks of Ralph Waldo Emerson*, vol. 10, ed. Merton M. Sealts Jr. (Harvard University Press, 1973).
7. Roderick Nash, *Wilderness and the American Mind*, 3th ed. (Yale University press, 1967).
8. Nicholas L. Guardiano, *Aesthetic Transcendentalism in Emerson, Peirce, and Nineteenth-Century American Landscape Painting* (Lexington Books, 2017).
9. Immanuel Kant, *Critique of Judgement*, trans. Werner S. Pluhar (Hackett Publishing, 1790/1987).
10. Emerson, *The Journals and Miscellaneous Notebooks of Ralph Waldo Emerson*, vol. 16. Ed. Ronald A. Bosco and Glen M. Johnson (Harvard University Press, 1982).
11. Donald Worster, *A Passion for Nature: The Life of John Muir* (Oxford University Press, 2008).
12. Henry David Thoreau, *Walden or Life in the Woods and Civil Disobedience* (Signet Classics, 1999).
13. Wilderness Act, Public Law 88-577, 88th Congress, 1964, Sec. 2 (c).
14. John Copeland Nagle, "The Spiritual Values of Wilderness, "35 Envtl. L. 955 (2005).
15. Holmes Rolston III, *Philosophy Gone Wild* (Prometheus Books, 1989).
16. Aldo Leopold, *A Sand County Almanac* (Oxford University Press, 1949).

Chapter 16
Chinese Philosophy and Nature as a Place

In Chinese culture, there has been a very long tradition of aesthetic appreciation of nature, expressed in Chinese landscape paintings, classical literature, poems, Chinese classical gardens such as Suzhou gardens, music, dancing, architecture, and food. The Chinese aesthetic appreciation of nature is usually directed towards places such as farmland, gardens, mountains and waters, and birds. In public discourse, people usually use various concepts such as mountains and waters, mountains and forests, farmland, and gardens to express the concept of nature. Aesthetic appreciation of nature in Chinese culture is deeply influenced by Chinese philosophy, especially Confucianism, Daoism, and Zen Buddhism. In this book, I only focus on Confucianism and Daoism. In the following, I will examine two philosophical questions: (1) What is the metaphysical foundation for aesthetic appreciation of nature in China? (2) How do nature aesthetics in China promote the aesthetic appreciation of cultivated places while neglecting wilderness?

1 Metaphysical Foundation for Aesthetic Appreciation of Nature and the Expression of Material Nature

In part II of this book, I have examined the metaphysics in Confucianism and Daoism. In this section, I will explore how metaphysics has a direct influence on the aesthetic appreciation of nature. The concept of Heaven and Earth in Chinese philosophy is deeply connected with the value of goodness, which is not transcendent like Emerson understands. Rather it is derived directly from social values. The basic values of Confucianism and Daoism are in conflict. Confucianism focuses on the moral dimension of human beings. It stresses that human beings become complete only if they are devoted to the moral cultivation of the core virtues which are benevolence, righteousness, propriety, wisdom, and trust. Daoism stresses the virtue of wandering

S. Gao, *Comparative Studies Between Emerson and Chinese Philosophy*, Nature Aesthetics: Interdisciplinary and Intercultural Dialogue,
https://doi.org/10.1007/978-981-95-0421-3_16

mind which does not place value on morality alone; rather it gives equal consideration to all other forms of value in life. However, one basic value shared by both Confucianism and Daoism is living in harmony with nature. The Chinese expression for this is 天人合一, which is often translated as "unity between human beings and nature". In Chinese philosophy, aesthetic appreciation of nature is closely related to the good life. For both Confucianism and Daoism, the good life is virtuous life. This central point is important for us to understand the Chinese aesthetic appreciation of nature.

Chinese philosophy pursues a good life though living in harmony with nature. In the philosophical systems of Confucianism and Daoism, how to live a good life is an eternal theme, but in their thinking life itself is the metaphysical field and they usually do not focus on an afterlife or a transcendent existence with God, as Western philosophy and religion usually do. Spinoza and Aristotle, for example, both view a good life as one which is related in the right way to God. For Spinoza, the good life is characterized by knowing God since God is considered as the origin of truth. For Aristotle, the best life is not related to everyday practical life; rather it is the contemplative life, which is a life with God. Aristotle's concept of God is not concerned with human affairs or happiness. By comparison, in Confucianism and Daoism, there is no transcendental God. The ultimate reality is Dao, which is immanent in nature and everyday life, and the good life is to live in harmony with Dao. Since Dao is imminent in nature, aesthetic appreciation of nature is an important part of living a good life.

Then how do Chinese philosophy convey the concept of material nature or physical nature? Material nature in Confucianism and Daoism is referred to as "the myriad things" (万物). Heaven and Earth are often understood as the opposing powers of *yin* and *yang* which generate myriad things. The aesthetic appreciation of myriad things is based on their relationship with human beings' virtue. When their external features fit with the human virtues, aesthetic appreciation of things occurs. For example, in Chinese culture, bamboo is aesthetically appreciated since the shape of bamboo is hollow inside and it is upright outside. It represents people's pursuit of virtue and cultivation of virtue.

In the classical texts of Confucianism and Daoism, material nature serves as a moral symbol rather than being treated as having its metaphysical status. The aesthetic appreciation of material nature is in essence the appreciation of the values which are interpreted by philosophers. The basic value of material nature lies in its harmonious relationship with human beings, a relationship constituting the traditional basis for Chinese civilization. People's living directly depends on agriculture for survival, which created a rich agrarian culture. Material nature is thus placed in the living world of human beings and people at that time seldom observe and appreciate nature for its own sake, as some modern scientists and artists have done. In Confucian and Daoist texts, typically every mention of the concept of nature will be followed by mention of virtues of human beings.

If individual items in nature are not related to the harmonious relationship between human beings and nature, they will not be given aesthetic attention and appreciation. For example, the great forces of nature such as flooding, earthquakes, tornadoes and so on are not aesthetically appreciated since they will bring disaster for agricultural production and people's everyday lives.

In terms of material nature, there are two elements which are frequently mentioned in the Classical text of Chinese philosophy: mountains and waters (山水). The two combined words are often used in daily conversation and institutional name and official documents of environmental policy. Let us first take a look at the classical text of Confucianism: *The Analects*. The Master said, "The wise find joy in water; the benevolent find joy in mountains. The wise are active; the benevolent are still. The wise are joyful; the benevolent are long-lived."[1] "Mountain" and "Water" are often used together in Chinese to express aesthetic appreciation of nature. One reason is that geographically, mountains and water dominate the entire Chinese landscape. The other reason is that Confucian thought associates moral values with them based on their shape and texture. The shape and texture of mountains represents stability and reliability, which is the key feature of the dominant Confucian virtue of "benevolence". Since benevolence is associated with responsibility, the heaviness of mountains thus signifies responsibility. In the book of *Mencius*, mountains are also often mentioned.

> The trees of the Niu Mountain were once beautiful. But can the mountain be regarded any longer as beautiful since, being in the borders of a big state, the trees have been hewed down with axes and hatchets? Still with the rest given them by the days and nights and the nourishment provided them by the rains and the dew, they were not without buds and sprouts springing forth. But then the cattle and the sheep pastured upon them one and again. That is why the mountain looks so bald. When people see that it is so bald, they think that there was never any timber on the mountain. Is this the true nature of mountain? Is there not [also] a heart of humanity and righteousness originally existing in man? [2]

For Confucianism, the mountain has its own nature, which is characterized by supporting life around it. That is, Confucianism uses the mountain as symbol for the expression of the Confucian core virtue of benevolence. Benevolence is characterized by loving and supporting lives, especially those of human beings. Mencius does not have aesthetic appreciation of mountain and trees based on their own features, as examined in modern ecology, biology, and geology. Rather the mountain is used as a symbol for expressing human virtues which are the major object of aesthetic appreciation. In Zhuangzi's work, the concept of the mountain also frequently appears. "He said that there is a Holy Man living on faraway Ku-she Mountain, with skin like ice or snow, and gentle and shy like a young girl. …By concentrating his spirit, he can protect creatures from sickness and plague and make the harvest plentiful."[3] When Zhuangzi mentions mountains in his work, he will not focus on the objective features of mountains. Instead, mountains for him are the dwelling place for virtuous person, which is often termed as authentic person. (真人)Zhuangzi usually focuses

---

[1] Confucius, *The Analects*, trans. D. C. Lau (Penguin Books, 1979), p. 84.

[2] *A Source Book in Chinese Philosophy*, trans. Wing-tsit Chan (Princeton University Press, 1963), p. 56.

[3] Zhuangzi, *The Complete Works of Chuang Tzu*, trans. Burton Waston (New York: Columbia University Press, 1968), p. 33.

on vast space in the mountains which is the symbol for him to express the infinity of Dao. In terms of waters, Confucianism and Daoism both aesthetically appreciate them for their connection with their ideal virtues. In Zhuangzi, there is a similar passage regarding water. "When water is perfectly still, it can serve as a standard of levelness. When the internal is preserved, the external will remain unshaken as well. Virtuosity is really just an ornamental form giving a definite shape to the fullness of the harmony within."[4] From the two passages, it can be seen clearly that Confucius and Zhuangzi both appreciate water for their different interpretation of virtues. Aesthetic appreciation of water is based on this interpretation. They will not focus on water itself and appreciate for its own sake.

In addition to mountains and waters, many natural elements such as plants and animals also frequently appear in the classical texts of Chinese philosophy. However, they are used only as symbols to convey the idea of a virtuous life. These elements themselves receive little attention. One passage from Zhuangzi can best illustrate this point.

> Chuang Chou was wandering in the park at Tiao-ling when he saw a peculiar kind of magpie that came flying along from the south. It had a wingspread of seven feet and its eyes were a good inch in diameter. It brushed against Chuang Chou's forehead and then settled down in a grove of chestnut trees. "What kind of bird is that!" exclaimed Chuang Chou. "Its wings are enormous but they get it nowhere; its eyes are huge but it can't even see where it's going!" Then he hitched up his roble, strode forward, cocked his crossbow and prepared to take aim. As he did so, he spied a cicada that had found a lovely spot of shade and had forgotten all about [the possibility of danger to] its body. Behind it, a praying mantis, stretching forth its claws, prepared to snatch the cicada, and it too had forgotten about its own form as it eyed its prize. The peculiar magpie was close behind, ready to make off with the praying mantis, forgetting its own true self as it fixed its eyes on the prospect of gain. Chuang Chou, shuddering at the sight, said, "Ah!— things do nothing but make trouble for each other—one creature calling down disaster on another!" He threw down his crossbow, turned about, and hurried from the park, but the park keeper [taking him for a poacher] raced after him with shouts of accussion.[5]

From this passage, we can see that Zhuangzi has a close observation towards nature and he discovers the food chain in the system of nature. However, his attention is not on deep understanding of the food chain in nature, rather to find moral lesson from nature. The moral lesson from observation of nature will guide him reflect on his self and a good life. The core value Zhuangzi stresses is *wuwei* (effortless action or no action). It is a human value, which is projected into the value of Heaven and Earth, which will in turn influence human beings' action. By comparison, Emerson also gives a close observation of nature by the help of developing scientific knowledge in his lifetime. He is full of aesthetic appreciation of nature itself which is guided by both scientific knowledge and poetic imagination. He writes, "Nature, in its ministry to man, is not only the material, but is also the process and the result. All the parts incessantly work into each other's hands for the profit of man. The wind sows the

[4] Zhuangzi, *Zhuangzi: The Essential Writings with Selections from Traditional Commentaries*, trans. Brook Ziporyn (Hackett Publishing Company, 2009), p. 37.

[5] Zhuangzi, *The Complete Works of Chuang Tzu*, trans. Burton Waston (New York: Columbia University Press, 1968), pp. 218–219.

seed; the sun evaporates the sea; the wind blows the vapor to the field; the ice, on the other side of the planet, condenses rain on this; the rain feeds the plant; the plant feeds the animal; and thus the endless circulations of the divine charity nourish man."[6] "For example, in the animal kingdom the physiologist has observed that no creatures are favorites, but a certain compensation balances every gift and every defect. A surplusage given to one part is paid out of a reduction from another part of the same creature."[7] Emerson's observation of nature is directly influenced by science. He discovered the truth in nature. Truth for him is spiritual law. Emerson is full of love towards the spiritual law and that is the metaphysical foundation for his strong affections towards all kinds of nature. For him, farming land is beautiful but there is more harmony in wilderness itself.

2 Chinese Philosophy and Sense of Place

The metaphysical interpretation of Heaven and Earth in Chinese philosophy supports the sense of place. Material nature in Chinese philosophy is not explained through mathematics, geometry or physics, rather its existence as the world of life in human beings is taken as the self-evident truth. The world of life is the Chinese ontology. Dao is immanent in this life and this world is full of meaning and value. The Chinese aesthetic appreciation of mountains, waters, plants, and agricultural land is based on this ontology of everyday life. The Chinese aesthetic appreciation of nature can be characterized by "everyday aesthetics".

Professor Yuedi Liu, who raised the concept of "everyday life aesthetics" examines this concept in his book titled *Living Aesthetics: Critique of Modernity and Reconstruction of Aesthetic Spirituality*.[8] It is an aesthetic that is rooted in daily life and encompasses various aspects related to agriculture, including food, agricultural seasons, spatial elements associated with agriculture, and sustainable ways of life that emerge from agricultural practices. The basic point is that the Chinese aesthetic appreciation of nature is deeply ingrained in the agrarian tradition.

The aesthetic appreciation of wilderness requires a departure from the everyday aesthetic style of appreciation, since wilderness often poses a danger to human beings. However, such as perspective is not adopted in Zhuangzi's philosophy. Rather, when wilderness is considered, it is reconceived as a space largely detached from the everyday experiences of human life. It is outside the agricultural land, which is the main living space for people. This can be seen in Zhuangzi's concept of the "World of Supreme Virtue" (至德之世).

---

[6] Ralph Wado Emerson, *The Portable Emerson* , ed. Jeffrey S. Cramer (Penguin Books, 2014), p. 21.

[7] Ibid., p. 176.

[8] Yuedi Liu, *Living Aesthetics: Critical Modernity and Reconstruction of Aesthetics* (Beijing Normal University Press, 2024).

> For the people too have their own constant inborn nature. To be clothed by their own weaving, fed by their own plowing—this is what is called their shared Virtuosity. All as one, without faction—I call that simply the way Heaven has cast them forth. Thus, in the age of perfect Virtuosity their actions were solid and full but their gaze was distant and blank. For in those days, there were no paths or trails through the mountains, no boats or bridges over the ponds; all creatures lived together, merging their territory into one another. The birds and beasts clustered with each other, the grasses and trees grew unhampered, so one could tie a cord to a bird or beast and take a stroll with it or bend down a branch to peep into a bird's nest. Indeed, in those days of perfect Virtuosity, the people lived together with the birds and beasts, bunched together with all things.[9]

In the above passage, we can find that there are two types of landscapes: farming land and wilderness beside the farming land. In contrast to Emerson's appreciation of wilderness, Zhuangzi views farmland as the central domain of human existence. He pays little attention to the rich biodiversity of untamed wilderness. Rather, his focus lies on cultivated farmland. Zhuangzi has the deep aesthetic appreciation for this cultivated place, as it symbolizes and facilitates the cultivation of supreme virtue through a life in harmony with nature. A question may arise here: why is farming land related to virtue? Farming land is deeply connected to the activities of labor and the production of food on the land. Zhuangzi thinks that these agricultural activities are demonstrations of the essence of human nature. His conception of place differs greatly from Emerson's. For Emerson, our self is related to God, rather than farming activities relating to the land. I want to use one passage from Emerson's work to conduct a comparative study in their understanding of place. "All places are alike to him for in all appear that law and order which certify him of co-nature with his own constitution more than do the lineaments which resemble his own Home in that which affirms itself to be the Cause of all, all his knowledge and moral growth go to domesticate him in every fact and event that transpire in nature. All places are alike to him for that transpire in nature. All places are alike to him for that which is with him constitutes places."[10] For Emerson, the real home for him is the final cause of all things in nature, which is also termed as spiritual law. All of the places are the production of spiritual law and they are beautiful. But the beauty is inferior to the beauty of God. For him, physical things are the best interpreters of God.

Emerson is full of aesthetic appreciation for all creatures in this world. However, among various types of landscapes such as farming land, gardens, and wilderness, Emerson appreciates wilderness more than his appreciation of gardens and farmland. This point can be demonstrated by the following passages from his essay "Nature", "The charming landscape which I saw this morning is indubitably made up of some twenty or thirty farms. Miller owns this field, Locke that, and Manning the woodland beyond. But none of them owns the landscape. There is a property in the horizon which no man has but he whose eye can integrate all the parts, that is, the poet. This is the best part of these men's farms, yet to this their warranty-deeds

---

[9] Zhuangzi, *Zhuangzi: The Essential Writings with Selections from Traditional Commentaries* (Hackett Publishing Company, 2009), p. 61.

[10] *The Early Lectures of Ralph Waldo Emerson.* ed. Stephen E.Whicher, Robert E. Spiller, and Wallace E. Williams. Vol 3 (Cambridge, MA: Harvard University Press, 1959–1972), pp. 28–29.

give no title."[11] "Within these plantations of God, a decorum and sanctity reign, a perennial festival is dressed, and the guest sees not how he should tire of them in a thousand years. In the woods, we return to reason and faith."[12]

From the two passages about nature, we can see that farmland for Emerson only represents part of nature. Although it is beautiful, the beauty is inferior to the beauty of the whole nature. The woods represent the pristine nature and it is more beautiful than the farmland. When human beings return to the woods, they can get better moral guidance. The representative person who can best appreciate the beauty of nature is poet rather than farmers. He writes, "Yet this may show us what discord is between man and nature, for you cannot freely admire a noble landscape if laborers are digging in the field hard by. The poet finds something ridiculous in his delight until he is out of the sight of men."[13]

The question I would like to discuss here is why Emerson loves the wilderness more? I think the answer can be found in his concept of virtue. As I have examined before, Emerson stresses the cognitive powers and development in the mind. When we advance these powers, we will gain virtue. Wilderness is full of a rich biodiversity of life and it can best train the cognitive powers in our minds.

In his lecture on Country Life (1858), he talks a lot about the art of walking. For him, when we walking, it is good to bring two companions,"One is artist, or one who has an eye for beauty. ...The other is a naturalist, for the reason that it is much better to learn the elements of geology, of botany, or ornithology, and astronomy, by word of mouth from a companion, than drily by book."[14] From the above passage, we can find that Emerson stresses developing cognitive abilities in our mind. Compared with Emerson, Chinese Philosophy tends to neglect aesthetic appreciation wilderness. In China, wilderness has long be seen as a meaningless space, which doesn't have any value. Based on the explanation of Chinese dictionary, we can see that the Chinese interpretation of wilderness is characterized by three features: (1)It is closely related to the west region of China, and in ancient China, wilderness was often used as space where criminals were exiled. In wilderness, few people live there and life is very difficult; (2)The Chinese conception of wilderness is often associated with war, and the places where wars ended will become wilderness; (3)Wilderness is also associated with uncivilized places, representing border areas. However, in recent years, due to the construction of national park system in China, the emergence of various outdoor sports, and the environmental education activities in wilderness, wilderness documentaries and various media coverage, people's attitude towards wilderness has gradually begun to change.

A few philosophers in China have absorbed Rolston's concept of intrinsic value, trying to change the perception of wilderness in Chinese culture. But due to the influence of Chinese philosophy, the idea in the concept of intrinsic value in wilderness

[11] Ralph Waldo Emerson, *The Portable Emerson* (Penguin Books, Jeffrey S. Cramer, 2014), p. 19.

[12] Ibid., p. 20.

[13] Ibid., p. 48.

[14] Ralph Waldo Emerson, *The Best Read Naturalist*, ed. Michael P. Branch and Clinton Mohs (University of Virginia Press, 2017), pp. 201–202.

cannot convince people. I pointed out in my article titled "Can Chinese philosophy Embrace Wilderness?"[15] that the idea in the concept of intrinsic value is a product of Western philosophical metaphysics, epistemology, and ethical framework, and it does not fit with the core spirit of unity of Heaven and Man in Chinese philosophy. We should start from our own philosophical traditions and establish the new conception of unity between Heaven and Man in the modern context. Question may arise here: Why don't Heaven and Earth in Chinese philosophy include the wilderness? Confucianism and Daoism are in essence human centered. They both focus on living a virtuous life. In the past, wilderness is full of danger due to large animals and it is very difficult to get access to the space. Agriculture in the past is the core economic foundation of China. The temporal and spatial continuity in farming has greatly shaped the Confucian relationships. The emphasis on the virtue of filial piety and benevolence is closely related to this type of economic form. Stress on the virtues among human relationships also pose a challenge for people to transcend themselves to establish a new relationship between wilderness and human beings. However, in modern times, due to rapid process of urbanization and the emergence of virtual spaces, people will experience novel relationships which have the tendency to reconstruct the traditional self. Reconstruction of self is also a process of creating new values. I will discuss it in the last Chapter.

References

1. *A Source Book in Chinese Philosophy*, trans. Wing-tsit, Chan (Princeton University Press, 1963).
2. Zhuangzi, *Zhuangzi: The Essential Writings with Selections from Traditional Commentaries*, tras. Brook Ziporyn (Hackett Publishing Company, 2009).
3. Yuedi Liu, *Living Aesthetics: Critical Modernity and Reconstruction of Aesthetics* (Beijing Normal University Press, 2024).
4. Ralph Waldo Emerson, *The Portable Emerson*, ed. Jeffrey S. Cramer (Penguin Books, 2014).
5. Ralph Waldo Emerson, *The Best Read Naturalist*, ed. Michael P. Branch and Clinton Mohs (University of Virginia Press, 2017).

[15] Shan Gao, "Can Chinese Philosophy EmbraceWilderness?" Environmental Ethics 39(2), pp.147–159 (2017).

Chapter 17
Self-transcendence: Ecological Aesthetic Self and Environmental Virtue

Through an intercultural dialogue between Emerson and Chinese philosophy from the concepts of nature, spirituality, and place, the final purpose of this book is to establish self-transcendence. The essence of self-transcendence is to achieve the transition from virtuous self to ecological aesthetic self, which can promote the new virtue which is environmental virtue. To achieve this type of self, it needs self-transcendence. John Nolt in his article "Hope, Self-Transcendence and Environmental Ethics"[1] provides a similar approach by linking environmental virtue to the meaning of life through the concept of self-transcendence. However, I will modify this approach by relating environmental virtue with ecological aesthetic self, value in nature, and power.

Through the analysis of the concept of Heaven and Earth and virtue in Confucianism and Daoism, we can see that although living in harmony with nature is the spiritual pursuit, material nature serves as a symbol for human beings to cultivate virtue. It is human-centered rather than nature-centered. In the modern context of China, with the deepening understanding of nature and the establishment of national parks, our sense of wonder toward nature has been increasingly heightened. This sense of wonder forms the basis for recognizing nature's value, providing a foundation for environmental virtue that integrates the self with value. The development of science and the popularization of science education in China enable us to incorporate Emerson's nature-based aesthetics, grounded in natural science, into new virtue: environmental virtue. Below, I elaborate on these three elements: (1) ecological aesthetic self; (2) value in nature; (3) environmental virtue.

First, the ecological aesthetic self is grounded in natural science and the immersive bodily experience of being in nature. Although Confucianism-Daoism texts contain many descriptions of nature, these descriptions serve as symbols of the Dao. In Chinese philosophy, there is no aesthetic appreciation of nature itself, which is closely tied to their value system. Confucianism-Daoism's aesthetics revolve around

[1] John Nolt , "Hope, Self-Transcendence and Environmental Ethics", *Inquiry*, 53(2), p. 167.

S. Gao, *Comparative Studies Between Emerson and Chinese Philosophy*, Nature Aesthetics: Interdisciplinary and Intercultural Dialogue,
https://doi.org/10.1007/978-981-95-0421-3_17

human beings, focusing on a human-centered aesthetic and the object of beauty is harmony between humans and nature. The core of Confucianism-Daoism human-nature relationship lies in the harmony derived from the sustainable use of land. The ecological aesthetic self I stressed in this chapter expands the boundaries of this agrarian aesthetic to include the aesthetic appreciation of wilderness. However, their argument is problematic itself. Some scholars argue that Daoist philosopher Zhuangzi exhibits aesthetic appreciation of the wilderness. In Zhuangzi's philosophy, the wilderness represents a vast spatiality, such as mountains and forests, which he believed was conducive to cultivating the human spirit. However, metaphysically, Zhuangzi's philosophy doesn't support aesthetic appreciation of all aspects of wilderness.

In Zhuangzi's texts, all descriptions of nature are grounded in the transcendent value of the Dao, serving as expressions of the Dao rather than depictions of natural elements themselves. This differs significantly from Emerson, who, grounded in natural science, describes and appreciates natural objects themselves, as they embody the spiritual laws of God. The aesthetic appreciation of nature reflects a love for these laws. In Zhuangzi, however, the wilderness is not appreciated for itself as a physical or real entity but is instead presented as an imagined place. The natural elements he mentions serve merely as symbolic expressions within this imagined place. Both Confucianism and Daoism neglect wilderness appreciation. In terms of Confucianism's neglect of wilderness appreciation, I have given a deep examination in my paper titled Can Chinese Philosophy Embrace Wilderness?[2] I argue that Confucianism and Daoism's living in harmony with nature promotes the aesthetic appreciation of cultivated land, which has unique features combining time, space, direction, and language and so on.

Second, the value of nature. Nature's value is the result of the development of ecological aesthetic self. In Chinese philosophy, nature or wilderness itself doesn't have independent intrinsic value; rather its relationship with human beings has value. In Emerson's philosophy, the recognition of nature's value is achieved through God, who created nature, endowing it with a miraculous secondary value. In the Chinese cultural context, without the concept of God, recognizing nature's value can be achieved by cultivating a sense of wonder, which establishes a connection between humans and nature's value. This aesthetic sense of wonder arises from the combination of scientific understanding and immersive experiences in nature, while modern technological advancements allow people to access previously inaccessible landscapes without safety concerns. The logical relationship between wonder and nature's value has been extensively studied, with a consensus that wonder facilitates a non-anthropocentric perspective, encouraging concern for the well-being of others. For example, American philosopher Martha C. Nussbaum explains the relationship between wonder and nature's value as follows: "This emotion responds to the pull of the object, and one might say that in it the subject is maximally aware of the value of the object, and only minimally aware, if at all, of its relationship to her own plans.

[2] Shan Gao, "Can Chinese Philosophy Embrace Wilderness?", *Environmental Ethics* (Summer 2017, Volume 39, Number 2), pp. 147–159.

That is why it is likely to issue in contemplation, rather than in any other sort of action toward the object."[3] Similarly, Australian scholar Haydn George Washington, in his book title *A Sense of Wonder towards Nature: Healing the Planet through Belonging* defines wonder as follows: "A sense of wonder is transformative to us belonging in life and allows us to find the well-being of humanity as a part of the well-being of nature, the nature we feel wonder at, respect, and feel a duty to protect."[4] Through their definitions and explanations, we can see the deep connection between wonder and intrinsic value. However, this sense of wonder requires cultivation to become a stable emotion rather than fleeting curiosity. In Emerson's philosophy, this wonder is primarily achieved through faith, as a response to the creations of God. In the context of Chinese cultural traditions, if Zhuangzi's thought is to be modernized, the cultivation of wonder toward nature must be achieved through immersive experiences, scientific understanding, and narratives about nature that spark imagination.

Third, ecological aesthetic self can promote a new virtue, which is termed as environmental virtue. By environmental virtue, I refer to the duty to appreciate and protect the environment. It can motivate our aesthetic obligation to appreciate different types of nature including agriculture, gardens, and wilderness. Environmental virtue in essence is a transcendence. It transcends the traditional virtue which is constructed by dominating values in a certain culture. Both Emerson and Chinese philosophy have deep influence on the construction of virtuous self in their culture. However, this construction will be transcended by new values. Aesthetic appreciation of nature can help people transcend the old values and embrace new values. In this aspect, John Muir's transformation from virtue based on traditional Christianity to new environmental virtue which is directly motivated by aesthetic appreciation of nature can give us a deep insight. Donald Worster, the biographer of John Muir points out that Muir overcomes the traditional values which are shaped by traditional Christianity and finds value in wilderness. For him, wilderness becomes his new God. He transcends the virtuous self which is shaped by traditional Christianity and develops a new ecological aesthetic self which reshapes his virtue. Through his aesthetic appreciation of nature, Muir gained spiritual strength, enabling him to transcend the self shaped by the traditional religious values of his family and society. For instance, Donald Worster writes in Muir's biography, "Muir's had walked away from his father's Bible fundamentalism, which demanded chapter and verse for every belief. He was losing faith in the Christian story of a supernatural messiah come down from Heaven to live and die among mortals and offer them redemption in another world. But he had not lost his capacity for reverence. Nature had come to supplant written revelation as a source of truth and as the only heaven he needed, and before that nature, an old-fashioned humble, unquestioning reverence remained the only

[3] Martha C. Nussbaum, U*pheavals of Thought: The Intelligence of Emotions* (Cambridge University Press, 2001), p. 54.

[4] Haydan Washington *A Sense of Wonder Towards Nature: Healing the Planet through Belonging* (London: Routledge, 2019), p. 188.

acceptable stance."[5] Clearly, through aesthetic appreciation, Muir transcended the moral self shaped by traditional Calvinism, developing a wilderness aesthetic self. It is important to emphasize that this transcendence does not involve a complete abandonment of the previous self but rather the construction of a new self. This new self retains the core of the original spirit while incorporating new dimensions.

Environmental virtue can promote new types of ecological practices in China. Right now, China has launched the National Park project. However, the concept of wilderness has been neglected due to traditional cultural values. The new virtue can promote wilderness protection in China. Scholars also have raised the idea that we can promote wilderness conservation in China by using the opportunity of constructing national park system as I mentioned in part I. Some non-profit organizations in China have been promoting this for the past 50 years. However, without the participation of general public, wilderness conservation ideal can not be achieved. If the new virtue is formed, the general public can increasingly pay attention to wilderness conservation in the future.

References

1. John Nolt, "Hope, Self-Transcendence and Environmental Ethics", *Inquiry*, 53(2).
2. Shan Gao, "Can Chinese Philosophy Embrace Wilderness?", *Environmental Ethics* (Summer 2017, Volume 39, Number 2).
3. Martha C. Nussbaum, *Upheavals of Thought: The Intelligence of Emotions* (Cambridge University Press, 2001).
4. Haydn Washington *A Sense of Wonder Towards Nature: Healing the Planet through Belonging* (London and New York: Routledge, 2019).
5. Donald Worster, *A Passion for Nature: The Life of John Muir* (Oxford University Press, 2008).

[5] Donald, Worster, *A Passion for Nature: The Life of John Muir* (Oxford University Press, 2008), p. 142.

Conclusion

In this book, I conduct comparative studies between Emerson and Chinese philosophy from the concepts of nature, spirituality, and place. The traditional approach to comparative analysis between Emerson and Chinese philosophy tends to focus on the core philosophical concepts such as transformation, virtue, nature, or self-cultivation. In recent years, some scholars choose the ecological approach to this comparative study in response to the global environmental crisis. My book takes the alternative approach of spirituality. By spirituality, I mean people's pursuit of meaning in life through transcending the false self to the true self towards the ultimvate value of goondess. Emerson and Chinese philosophers (mainly Confucius, Mencius, and Zhuangzi) all pursue transcending the false self to the virtuous self through a life of living in harmony with nature. This is the unique feature of their spirituality, which has a profound impact on their experience of physical and material space. Space is no longer understood as geographical, or geometrical existence, rather, it becomes a place for human beings. Emerson's spirituality generates an aesthetic revolution. Nature, especially wilderness becomes the object of aesthetic appreciation. Aesthetic appreciation of wilderness promotes love towards wilderness so that wilderness becomes a place for Emerson. In contrast, Confucianism and Daoism generate aesthetic appreciation of cultivated space such as gardens, agricultural land, and living space within town. Through intercultural dialogue between Emerson and Chinese philosophy, an ecological aesthetic self can be cultivated, which can generate a new environmental virtue. This ecological aesthetic self can develop new aesthetic horizons within each culture, thereby fostering a renewed sense of place: in China, wilderness itself can become a meaningful place, while in America, cultivated and agricultural landscapes can likewise be reinterpreted as places of deep significance. Such a new sense of place can provide the profound inner motivation for environmental protection in both countries. China may eventually establish its own wilderness laws while in America, various cultivated places such as agricultural land can gain the same importance as wilderness conservation.

S. Gao, *Comparative Studies Between Emerson and Chinese Philosophy*, Nature Aesthetics: Interdisciplinary and Intercultural Dialogue,
https://doi.org/10.1007/978-981-95-0421-3

References

Acampora, R. R. (1994). Using and abusing Nietzsche for environmental ethics. Environmental Ethics, 16(2), 187–194.
Alain, D. (2014). Aristotle and the dominion of nature. Environmental Ethics, 36(4), 411–428.
Ames, Roger T. (1998). Thinking from the Han: Self, Truth, and Transcendence in Chinese and Western Culture. State University of New York Press, Albany.
Ames, Roger T. and Rosemont, Henry Jr. (1999). The Analects of Confucius: A Philosophical Translation. Ballantine Books, New York.
Anderson, Q. (1971). The imperial self: An essay in American literary and cultural history. Knopf, New York.
Andrews, M. Barry (2017). "That Which Was Ecstasy Shall Become Daily Bread." Religions 8(4).
Ansell-Pearson, K. (2013). True to the Earth: Nietzsche's Epicurean care of self and world. In Hutter, H., & Friedland, E. (Eds.), Nietzsche's therapeutic teaching (pp. 97–116). Bloomsbury Academic, New York.
Aristotle (1924). Metaphysics, trans. W. D. Ross. Clarendon Press, Oxford.
Aristotle (1961). Parts of Animals, trans. A. L. Peck. Harvard University Press, Cambridge, MA.
Aristotle (1996). Physics, trans. Robin Waterfield. Oxford University Press, Oxford.
Aristotle. (2003). Metaphysics (Miao, L., Trans.). China People's Publishing House, Beijing.
Aristotle (2009). Nicomachean Ethics. Oxford University Press, Oxford.
Aristotle (2010). History of Animals, trans. Wu Shoupen. The Commercial Press, Beijing.
Aristotle. (2013a). Nicomachean ethics (Liao, S., Trans.). The Commercial Press, Beijing.
Aristotle. (2013b). Politics (Yan, Y., & Qin, D., Trans.). China Renmin University Press, Beijing.
Aristotle. (2016). Physics (Zhang, Z., Trans.). The Commercial Press, Beijing.
Augustine (1871). The City of God, Volume 1, ed. Marcus Dods. T&T Clark, Edinburgh.
Augustine (1991). Confessions, trans. Henry Chadwick. Oxford University Press.
Augustine. (2007). Confessions (Xu, L., Trans.). China Social Sciences Press, Beijing.
Augustine. (2010). On free will. Shanghai Century Publishing Group, Shanghai.
Augustine. (2018). City of God (Vols. I–II) (Wang, X., Trans.). People's Publishing House, Beijing.
Berkeley, I. (1999). The roots of Romanticism. Princeton University Press, Princeton.
Buber, M. (1991). Chinese tales: Zhuangzi, sayings and parables and Chinese ghost and love stories (Page, A., Trans.). Humanities Press, Atlantic Highlands.
Bruno, G. (1998). On cause, principle and the one (Tang, X., Trans.). The Commercial Press, Beijing.
Branch, Michael P. and Mohs, Clinton (eds) (2017). The Best Read Naturalist: Nature Writings of Ralph Waldo Emerson. University of Virginia Press.

S. Gao, *Comparative Studies Between Emerson and Chinese Philosophy*, Nature Aesthetics: Interdisciplinary and Intercultural Dialogue,
https://doi.org/10.1007/978-981-95-0421-3

Brown, Nahum and Franke, William (eds) (2016). Transcendence, Immanence, and Intercultural Philosophy. Palgrave Macmillan.

Bruno, Giordano (1998). Cause, Principle and Unity. Cambridge University Press.

Cabot, J. E. (2017). Emerson biography: Born for freedom (She, Z., Trans.). Heilongjiang Education Press, Harbin.

Cafaro, P. (2006). Thoreau's living ethics: Walden and the pursuit of virtue. University of Georgia Press, Athens.

Callicott, J. B. (1985). Intrinsic value, quantum theory, and environmental ethics. Environmental Ethics, 7(3), 257–275.

Carroll, Peter N. (1969). Puritanism and the Wilderness: The Intellectual Significance of the New England Frontier, 1629–1700. Columbia University Press.

Casey, Edward (1993). Getting Back into Place. Indiana University Press.

Casey, E. S. (1997). The fate of place: A philosophical history. University of California Press, Berkeley.

Chan, W.-T. (1963). A source book in Chinese philosophy. Princeton University Press, Princeton.

Chalmers, David J. (2022). Reality+: Virtual Worlds and the Problems of Philosophy. W. W. Norton & Company.

Chan, Wing-tsit (1963). A Source Book in Chinese Philosophy. Princeton University Press.

Chen, Y. (2019). The academic origins of mysticism. The Commercial Press, Beijing.

Chenyang, Li and Franklin P. (2015) Chinese Metaphysics and Its Problems. Cambridge University Press.

Chong, Kim-chong (2023). "Zhuangzi and the Issue of Human Nature." Dao 22.

Confucius (1979). The Analects, trans. D. C. Lau. Penguin Books.

Cottingham, J. (2005). The spiritual dimension: Religion, philosophy, and human value. Cambridge University Press, New York.

Crippen, M. (2023). Chinese thought and transcendentalism: Ecology, place and conservative radicalism. Religions. https://www.mdpi.com/journal/religions/special_issues/XOCHHPL3HM. Accessed 28 March 2025.

Curley, E. (1969). Spinoza's metaphysics: An essay in interpretation. Harvard University Press, Cambridge, MA.

David, S. (1991). Is Aristotle's teleology anthropocentric? Phronesis, 36(2), 179–196.

de Jonge, E. (2016). Spinoza and deep ecology. Routledge Taylor & Francis Group, London.

Del Caro, A. (2004). Grounding the Nietzsche rhetoric of Earth. Walter de Gruyter, Berlin.

Descartes, René (1996). Meditations on First Philosophy, trans. John Cottingham. Cambridge University Press.

Descartes, René (1998). Meditations on First Philosophy, trans. Donald A. Cress. Hackett Publishing.

Descartes, René (1998). Discourse on the Method and Meditations on First Philosophy, trans. Donald A. Cress. Hackett Publishing.

Descartes, R. (2004). Descartes' speculative philosophy (Shang, X., et al., Trans.). Jiuzhou Press, Beijing.

Descartes, R. (2014). First philosophy meditations (Pang, J., Trans.). The Commercial Press, Beijing.

Descartes, R. (2015). On the passions of the soul (Jia, J., Trans.). The Commercial Press, Beijing.

Descartes, R. (2016). First philosophy meditations (Wu, C., Trans.). Taiwan Strait Publishing House, Taipei.

Devall, B. (1988). Simple in means, rich in ends: Practising deep ecology. Green Print, London.

Devall, B., & Sessions, G. (1985). The development of natural resources and the integrity of nature. Environmental Ethics, 7(4), 293–322.

Dewey, J. (1925). Experience and nature. Open Court, Chicago.

Dewey, John (1929). Experience and Nature. W. W. Norton & Company.

Drenthen, M. (2009). Nietzsche and the paradox of environmental ethics. New Nietzsche Studies, 5(1/2), 12–25.

Ducharme, A. (2014). Aristotle and the dominion of nature. Environmental Ethics, 36(4), 411–428.
Emerson, Ralph Waldo (1860). The Conduct of Life. The Riverside Press.
Emerson, Ralph Waldo (1895). Lectures and Biographical Sketches. The Riverside Press.
Emerson, R. W. (1983). Emerson: Essays and lectures (Porte, J., Ed.). Literary Classics of the United States, New York.
Emerson, Ralph Waldo (1908). English Traits & Representative Men & Other Essays, ed. Ernest Rhys. J. M. Dent & Sons.
Emerson, Ralph Waldo (1909). Essays and English Traits. P. F. Collier.
Emerson, Ralph Waldo (1964). The Early Lectures of Ralph Waldo Emerson, vol. 2, eds. Stephen E. Whicher, Robert E. Spiller and Wallace E. Williams. Harvard University Press.
Emerson, Ralph Waldo (1971). The Collected Works of Ralph Waldo Emerson, vol. 1, ed. Alfred R. Ferguson. Harvard University Press.
Emerson, Ralph Waldo (1971). The Collected Works of Ralph Waldo Emerson, vol. 1, Nature, Addresses, and Lectures, ed. Robert E. Spiller and Alfred R. Ferguson. Harvard University Press.
Emerson, Ralph Waldo (1973). The Journals and Miscellaneous Notebooks of Ralph Waldo Emerson, vol. 10, ed. Merton M. Sealts Jr. Harvard University Press.
Emerson, Ralph Waldo (1982). The Journals and Miscellaneous Notebooks of Ralph Waldo Emerson, vol. 16, ed. Ronald A. Bosco and Glen M. Johnson. Harvard University Press.
Emerson, Ralph Waldo (1983). Essays & Poems, ed. Joel Porte. The Library of America.
Emerson, Ralph Waldo (1989–1992). The Complete Sermons of Ralph Waldo Emerson, ed. Albert J. von Frank et al. University of Missouri Press.
Emerson, Ralph Waldo (2001). The Later Lectures of Ralph Waldo Emerson, vol. 2, ed. Ronald A. Bosco and Joel Myerson. University of Georgia Press.
Emerson, Ralph Waldo (2003). The Collected Works of Ralph Waldo Emerson: Volume VI, The Conduct of Life, eds. Barbara L. Packer, Joseph Slater and Douglas Emory Wilson. Harvard University Press.
Emerson, Ralph Waldo (2011). The Journals and Miscellaneous Notebooks of Ralph Waldo Emerson, vol. 14, ed. William H. Gilman et al. Harvard University Press.
Emerson, Ralph Waldo (2014). The Portable Emerson, ed. Jeffrey S. Cramer. Penguin Books.
Emerson, Ralph Waldo (2017). The Best Read Naturalist, ed. Michael P. Branch and Clinton Mohs. University of Virginia Press.
Emerson, R. W. (2015a). Complete essays of Emerson (Pu, L., Trans.). Beijing Institute of Technology Press, Beijing.
Emerson, R. W. (2015b). On Nature·American scholars (Zhao, Y., Trans.). Life·Reading·New Knowledge Sanlian Bookstore, Beijing.
Emerson, R. W. (2016). The rules of life (He, Y., Trans.). Democracy and Construction Publishing House, Beijing.
Emerson, R. W. (2020). The beauty of reason (Sun, Y., Trans.). Hunan Literature and Art Publishing House, Changsha.
Emerson, R. W. (1960–1970). The journals and miscellaneous notebooks of Ralph Waldo Emerson (Vols. 1–16) (Gilman, W. H., et al., Eds.). The Belknap Press of Harvard University Press, Cambridge, MA.
Emerson, R. W. (1971–2013). The collected works of Ralph Waldo Emerson (Vols. 1–10) (Spiller, R. E., Ferguson, A. R., et al., Eds.). The Belknap Press of Harvard University Press, Cambridge, MA.
Emerson, R. W. (1959–1972). The early lectures of Ralph Waldo Emerson (Vols. 1–3) (Whicher, S. E., Spiller, R. E., & Williams, W. E., Eds.). The Belknap Press of Harvard University Press, Cambridge, MA.
Emerson, R. W. (1995). The later lectures of Ralph Waldo Emerson, 1843–1871 (Vols. 1–2) (Bosco, R. A., & Myerson, J., Eds.). Columbia University Press, New York.
Epicurus (1926). The Extant Remains, trans. Cyril Bailey. Clarendon Press.

Epicurus, & Lucretius. (2020). Nature and happiness: The philosophy of Epicurus (Bao, L., Liu, Y., & Wang, W., Trans.). China Social Sciences Press, Beijing.
Feng Youlan (1952). A History of Chinese Philosophy, trans. Derk Bodde. George Allen & Unwin.
Frey, Jennifer A. and Vogler, Candace (eds) (2018). Self-transcendence and Virtue: Perspectives from Philosophy, Psychology, and Theology. Routledge.
Frankl, Viktor E. (1996). "Self-transcendence As A Human Phenomenon", Journal of Humanistic Psychology.
Foster, S. E. (2002). Aristotle and the environment. Environmental Ethics, 24(4), 409–428.
Fox, W. (1986). Approaching deep ecology: A response to Richard Sylvan's critique of deep ecology. Board of Environmental Studies, University of Tasmania, Hobart.
Fox, W. (1990). Toward a transpersonal ecology: Developing new foundations for environmentalism. Shambhala, London.
Fox, W. (1984). Deep ecology: A new philosophy of our time? The Ecologist, 14(5/6), 194–200.
Gao, Shan (2017). "Can Chinese Philosophy Embrace Wilderness?" Environmental Ethics 39(2).
Gardner, Daniel K. (2007). The Four Books: The Basic Teachings of the Later Confucian Tradition. Hackett Publishing.
Great Books of the Western World, vol. 12, ed. Robert Maynard Hutchins (1952). Encyclopædia Britannica.
Guardiano, Nicholas L. (2017). Aesthetic Transcendentalism in Emerson, Peirce, and Nineteenth-Century American Landscape Painting. Lexington Books.
Guyton B. (1966). The Power of Self-transcendence: An Introduction to the Philosophical Theology of Paul Tillich. The Bethany Press.
Gaddi, L., et al. (1988). History of science, technology and philosophy in the 18th century (Vol. II) (Zheng, L., & Hu, J., Trans.). Zhejiang People's Publishing House, Hangzhou.
Goodman, R. B. (2004). The colors of the spirit: Emerson and Thoreau on nature and the self. In De Groot, J. (Ed.), Nature in American philosophy (pp. 1–18). Catholic University of America Press, Washington, DC.
Graham, A. C. (1986). The background of the Mencian theory of human nature. In Studies in Chinese philosophy and philosophical literature (pp. 63–78). Institute of East Asian Philosophies, Singapore (Re-published 1990, State University of New York Press, Albany).
Guardiano, N. L. (2011). Metamorphosis in art and nature: Emersonian poetry. Southwest Philosophical Studies, 33(1), 45–55.
Guardiano, N. L. (2016). Aesthetic transcendentalism in Emerson, Peirce, and nineteenth-century American landscape painting. Lexington Books, Lanham.
Hargrove, Eugene C. (1989). Foundations of Environmental Ethics. Environmental Ethics Books.
Hargrove, E. C. (2007). Foundations of environmental ethics (Yang, T., Jiang, Y., & Guo, H., Trans.). Chongqing Publishing House, Chongqing.
Harman, G. (1967). Toward a theory of intrinsic value. Journal of Philosophy, 64(23), 792–804.
Harvey, S. C. (2013). Transatlantic transcendentalism: Coleridge, Emerson, and nature. Edinburgh University Press, Edinburgh.
He, L. (1982). A brief description of Spinoza's philosophy. In Foreign philosophy (Vol. II) (pp. 45–60). The Commercial Press, Beijing.
He, L. (1986). The religious aspect of Spinoza's philosophy. Journal of the Graduate School of the Chinese Academy of Social Sciences, 1986(2), 23–35.
Heidegger, Martin (1984). Nietzsche, Volume II. Harper & Row.
Heidegger, M. (2018). Nietzsche (Sun, Z., Trans.). The Commercial Press, Beijing.
Hershock, P. D., & Ames, R. T. (Eds.). (2019). Philosophies of place: An intercultural conversation. University of Hawai'i Press, Honolulu.
Hill, T. E., Jr. (1983). Ideals of human excellence and preserving natural environments. Environmental Ethics, 5(3), 211–224.
Houle, K. L. F. (1997). Spinoza and ecology revisited. Environmental Ethics, 19(4), 417–431.
Huang, Y. (2013). On Spinoza's 'entity' theory and 'purposeful existence of man' thought. Journal of Central China Normal University (Humanities and Social Sciences), 52(4), 89–97.

James, W. (1925). The varieties of religious experience: A study in human nature. Longmans, Green, and Co., New York.

Jenkins (n.d.). Defending the Beauty of Nature: Aesthetic Obligations and the Establishment of Landscape Conservation Areas in England and Wales.

Jenkins, V. (2020). In defense of natural beauty: Aesthetic obligation and the law on the designation of protected landscapes in England and Wales. Environmental Law Review, 22(1), 45–62.

Jia, Jinhua (2021). "Writings, Emotions and Obligations: The Religious-Ritual Origin of the Classical Confucian Concept of Cheng (Sincerity)." Religions 12(6). https://doi.org/10.3390/rel12060382.

Kant, Immanuel (1956). Critique of Practical Reason, trans. Lewis White Beck. Bobbs-Merrill.

Kant, Immanuel (1965). Critique of Pure Reason, trans. Norman Kemp Smith. St. Martin's Press.

Kant, Immanuel (1987). Critique of Judgment, trans. Werner S. Pluhar. Hackett Publishing.

Kant, Immanuel (1996). The Metaphysics of Morals, trans. Mary Gregor. Cambridge University Press.

Kant, Immanuel (1997). Groundwork of the Metaphysics of Morals, trans. Mary Gregor. Cambridge University Press.

Kant, Immanuel (1998). Critique of Pure Reason, trans. Paul Guyer and Allen W. Wood. Cambridge University Press.

Kant, Immanuel (2000). Critique of Judgment, trans. Paul Guyer and Eric Matthews. Cambridge University Press.

Kant, Immanuel (2007). Critique of Pure Reason, trans. Marcus Weigelt. Cambridge University Press.

Kant, Immanuel (2011). Groundwork of the Metaphysics of Morals. Cambridge University Press.

Kant, Immanuel (2018). Groundwork of the Metaphysics of Morals. Yale University Press.

Kant, I. (2009). Kant's three major critiques collection (Vols. I–II) (Deng, X., Trans.). People's Publishing House, Beijing.

Kant, I. (2013a). Groundwork of the metaphysics of morals (Li, Q., Trans.). China Renmin University Press, Beijing.

Kant, I. (2013b). Metaphysics of morals (Zhang, R., & Li, Q., Trans.). China Renmin University Press, Beijing.

Konvitz, Milton R. (ed). (1972). The Recognition of Ralph Waldo Emerson: Selected Criticism Since 1837. University of Michigan Press.

Knudtson, P., & Suzuki, D. (1992). Wisdom of the elders. Allen & Unwin, Crows Nest, NSW.

Kraut, R. (2007). Flourishing: The central concept of practical thought. In Cairns, D., Herrmann, F.-G., & Penner, T. (Eds.), Pursuing the good: Ethics and metaphysics in Plato's Republic (pp. 45–67). Edinburgh University Press, Edinburgh.

Lafreniere, G. F. (1990). Rousseau and the European roots of environmentalism. Environmental History Review, 14(4), 41–72.

Lara, E. C., & Brouwer, R. (Eds.). (2017). Aristotle's practical philosophy: On the relationship between his ethics and politics. Springer, Cham.

Laozi (1963). Tao Te Ching, trans. D. C. Lau. The Chinese University Press.

Leopold, Aldo (1949). A Sand County Almanac. Oxford University Press.

Liu, C. (2018). Discussion on the system of Tian Xia in Chinese aesthetics. Exploration and Free Views, 2018(5), 78–89.

Liu, X. (2006). Laozi ancient and modern: Five kinds of collation and analysis introduction. China Social Sciences Press, Beijing.

Li Zehou (2008). The Origins of Chinese Thought: From Shamanism to Ritual Regulations and Humaneness, trans. Robert A. Carleo III. Brill.

Liu Yuedi (2024). Living Aesthetics: Critical Modernity and Reconstruction of Aesthetics. Beijing Normal University Press.

Lloyd, G. (1980). Spinoza's environmental ethics. Inquiry, 23(3), 293–311.

Malpas. J.E. (2004). Place and Experience: A Philosophical Topography. Cambridge University Press.

Masters, R. D. (2013). Rousseau's political philosophy (Hu, X., & Huang, T., Trans.). East China Normal University Press, Shanghai.

McPherson, D. (Ed.). (2017). Spirituality and the good life: Philosophical approaches. Cambridge University Press, Cambridge.

Mencius (1970). Mencius, trans. D. C. Lau. Penguin Books, London.

Mencius (2004). Mencius, trans. D. C. Lau. Penguin Books.

Mencius (2008). Mencius, trans. Bryan W. Van Norden. Hackett Publishing.

Meng Peiyuan (2009). Emotion and Reason. China Renmin University Press.

Metz, Thaddeus (2013). Meaning in Life. Oxford University Press, Oxford.Mintoff, Joe (n.d.) "Transcending Absurdity." Ratio 21(1).

Miller, E. (2012). Nietzsche on individuation and purposiveness in nature. In Bishop, P. (Ed.), A companion to Friedrich Nietzsche (pp. 57–74). Camden House, Rochester, NY.

Mirus, Christopher V. (2012). "Aristotle on Beauty and Goodness in Nature." International Philosophical Quarterly 52(1).

Myerson, Joel (2000). A Historical Guide to Ralph Waldo Emerson. Oxford University Press, Oxford.

Naess, Arne (1989). Ecology, Community and Lifestyle: Outline of an Ecosophy, trans. David Rothenberg. Cambridge University Press.

Naess, A. (1973). The shallow and the deep, long-range ecology movement. Inquiry, 16(1), 95–100.

Naess, A. (1974). Freedom, emotion and self-subsistence: The structure of a theoretical background. Universitetsforlaget, Oslo.

Naess, A. (1977). Spinoza and ecology. In Hessing, S. (Ed.), Speculum Spinozanum 1677–1977 (pp. 418–425). Routledge, London.

Naess, A. (1980). Environmental ethics and Spinoza's ethics: Comment on Genevieve Lloyd's article. Inquiry, 23(3), 313–325.

Nagle, John Copeland (2005). "The Spiritual Values of Wilderness." Environmental Law 35. Nash, Roderick (1967) Wilderness and the American Mind, 3rd ed. Yale University Press.

Nash, R. F. (2001). Wilderness and the American mind. Yale University Press, New Haven.

Nash, R. F. (2012). Wilderness and American thought (Hou, W., & Hou, J., Trans.). China Environmental Science Press, Beijing.

Nelson, E. S. (2011). The human and the inhuman: Ethics and religion in the Zhuangzi. Journal of Chinese Philosophy, 38(4), 589–610.

Nietzsche, Friedrich (1968). The Will to Power, ed. Walter Kaufmann. Vintage Books,.

Nietzsche, Friedrich (1974). The Gay Science, trans. Walter Kaufmann. Vintage Books.

Nietzsche, Friedrich (1983). Untimely Meditations, trans. R. J. Hollingdale. Cambridge University Press.

Nietzsche, Friedrich (1986). Human, All Too Human, trans. R. J. Hollingdale. Cambridge University Press.

Nietzsche, Friedrich (1989). Beyond Good and Evil, trans. Walter Kaufmann. Vintage Books.

Nietzsche, F. (2007). Untimely meditations (Li, Q., Trans.). East China Normal University Press, Shanghai.

Nietzsche, F. (2015a). On the genealogy of morality (Liang, X., Trans.). East China Normal University Press, Shanghai.

Nietzsche, F. (2015b). Twilight of the idols (Yang, D., & Chen, Y., Trans.). Jiangsu Phoenix Literature and Art Publishing House, Nanjing.

Nietzsche, F. (2017a). Beyond good and evil (Zhao, Q., Trans.). The Commercial Press, Beijing.

Nietzsche, F. (2017b). Human, all too human (Yang, H., Trans.). Renmin University of China Press, Beijing.

Nietzsche, F. (2017c). The gay science (Huang, M., Trans.). Lijiang Publishing House, Guilin.

Nietzsche, F. (2017d). Thus spoke Zarathustra. The Commercial Press, Beijing.

Nietzsche, F. (2018). The will to power (Sun, Z., Trans.). Shanghai People's Publishing House, Shanghai.

Nolt, J. (2010). "Hope, Self-Transcendence and Environmental Ethics." Inquiry 53(2).

Norton, B. G. (1984). Environmental ethics and weak anthropocentrism. Environmental Ethics, 6(2), 131–148.
Nussbaum, M., & Sen, A. (Eds.). (1993). The quality of life. Oxford University Press, New York.
Nussbaum, Martha C. (2001). Upheavals of Thought: The Intelligence of Emotions. Cambridge University Press.
Orff, G. (1963). History of science, technology and philosophy in the 18th century (Vol. II). The Commercial Press, Beijing.
Peng, G. (2019). Confucian tradition: Between religion and humanism. Peking University Press, Beijing.
Plato (1914). Euthyphro, Apology, Crito, Phaedo, Phaedrus, trans. Harold North Fowler. Harvard University Press.
Plato (1926). Laws, trans. R. G. Bury. Harvard University Press, Cambridge, MA. Plato (1968). The Republic, trans. Allan Bloom. Basic Books, New York.
Plato (1977). Phaedo, trans. G. M. A. Grube. Hackett Publishing, Indianapolis.
Plato (1982). Philebus, trans. Robin A. H. Waterfield. Penguin Books, London. Plato (1992). Republic, trans. G. M. A. Grube. Hackett Publishing, Indianapolis.
Plato (n.d.). The Republic, trans. Paul Shorey. Harvard University Press, Cambridge, MA. Plotinus (1991). The Enneads, trans. Stephen MacKenna. Penguin Books, London.
Plato. (2012). Phaedo (Yang, J., Trans.). China International Broadcasting Publishing House, Beijing.
Plato. (2015a). Complete works of Plato (Wang, X., Trans.). People's Publishing House, Beijing.
Plato. (2015b). The Republic (Guo, B., & Zhang, Z., Trans.). The Commercial Press, Beijing.
Plotinus. (2004). The Enneads. China Social Sciences Press, Beijing.
Plotinus. (2009). Enneads (Vols. I–II) (Shi, M., Trans.). China Social Sciences Press, Beijing.
Qian, M. (1996). Emerson and China: Reflection on individualism. Joint Publishing, Beijing.
Regan, Tom (1983). The Case for Animal Rights. University of California Press.
Richardson, Robert D. (1995). Emerson: The Mind on Fire. University of California Press.
Ripley, George (1882). George Ripley, by Octavius Brooks Frothingham. Houghton Miffli.
Robert Ford (2009). Making Transcendents: Ascetics and Social Memory in Early Medieval China. University of Hawai'i Press.
Rolston, Holmes (1988). Environmental Ethics: Duties to and Values in the Natural World. Temple University Press.
Rolston, Holmes III (1989). Philosophy Gone Wild. Prometheus Books.
Rolston, H., III. (1981). Is there an ecological ethic? Inquiry, 25(1), 81–93.
Rolston, H., III. (1986). Philosophy gone wild. Prometheus Books, Buffalo.
Rorty, R. (1980). Philosophy and the mirror of nature. Blackwell, Oxford.
Rousseau, Jean-Jacques (1973). The Social Contract and Discourses, trans. G. D. H. Cole. J. M. Dent.
Rousseau, Jean-Jacques (1979). Emile or On Education, trans. Allan Bloom. Basic Books.
Rousseau, Jean-Jacques (2004). Reveries of the Solitary Walker, trans. Peter France. Penguin Classics.
Rousseau, J.-J. (2013). Complete works of Rousseau (Li, P., Trans.). The Commercial Press, Beijing.
Rousseau, J.-J. (2015). Emile (Li, P., Trans.). The Commercial Press, Beijing.
Sedley, David (1991). "Is Aristotle's Teleology Anthropocentric?" Phronesis 36(2).
Skowroński, Krzysztof Piotr (2009). Values and Powers: Re-reading the Philosophical Tradition of American Pragmatism. Rodopi.
Schipper, K. (1994). The Taoist body. University of California Press, Berkeley.
Singer, P. (1993). Practical ethics. Cambridge University Press, New York.
Slingerland, E. (2003). Effortless action: Wu-wei as conceptual metaphor and spiritual ideal in early China. Oxford University Press, Oxford.
Sokolov, V. (1959). Spinoza's worldview. The Commercial Press, Beijing.
Solomon, R. C. (2002). Spirituality for the skeptic. Oxford University Press, Oxford.

Spinoza, B. (2014). Short treatise on God, man and his well-being (Vol. 1 of Spinoza's works). The Commercial Press, Beijing.
Spinoza, Baruch (1985). The Collected Works of Spinoza, trans. Edwin Curley. Princeton University Press.
Spinoza, Baruch (1994). Ethics, trans. Edwin Curley. Princeton University Press.
Spinoza, Baruch (1994). A Spinoza Reader: The Ethics and Other Works, trans. Edwin Curley. Princeton University Press.
Spinoza, B. (2016a). Ethics (Vol. 4 of Spinoza's works) (He, L., Trans.). The Commercial Press, Beijing.
Spinoza, B. (2016b). Principles of Cartesian philosophy (Vol. 2 of Spinoza's works) (Wang, Y., & Hong, H., Trans.). The Commercial Press, Beijing.
Spinoza, B. (2016c). The letters (Vol. 5 of Spinoza's works) (Hong, H., Trans.). The Commercial Press, Beijing.
Springer, I. (2010). Meaning in life: The harmony of nature and spirit. MIT Press, Cambridge, MA.
Stack, G. J. (1989). Emerson's influence on Nietzsche's concept of the will to power. Modern Schoolman, 66(3), 175–195.
Strauss, L. (2017). The natural and moral of the philosopher (Britz, Ed., Cao, C., Trans.). East China Normal University Press, Shanghai.
Tang, J. (1988). Collected essays on comparative study of Chinese and Western thoughts. Xuesheng Shuju, Taipei.
Tauber, A. I. (2012). "Thoreau": Moral epistemology and its contemporary relevance. Fordham University Press, New York.
Taylor, P. W. (1981). The ethics of respect for nature. Environmental Ethics, 3(3), 197–218.
Taylor, P. W. (1986). Respect for nature: A theory of environmental ethics. Princeton University Press, Princeton.
Taylor, P. W. (1992). The ethics of respect for nature. In Hargrove, E. C. (Ed.), The animal rights/environmental ethics debate (pp. 95–120). State University of New York Press, Albany.
The Bible. (2008). China Christian Council Publishing House, Beijing.
The Complete Works of Ralph Waldo Emerson, ed. Edward Waldo Emerson. Houghton Mifflin.
Thoreau, Henry David (1999). Walden or Life in the Woods and Civil Disobedience. Signet Classics.
Thompson, M., & Xu, L. (2013). A biocultural dialogue between Thoreau and Taoist thought: Rethinking environmental ethics, nature, spirituality and place. Religions. https://www.mdpi.com/journal/religions/special_issues/XOCHHPL3HM. Accessed 28 March 2025.
Thompson, P. B. (2010). The agrarian vision: Sustainability and environmental ethics. University Press of Kentucky, Lexington.
Thoreau, H. D. (1854). Walden; or, life in the woods. Ticknor and Fields, Boston.
Thoreau, H. D. (1997). Journal, vol. 5: 1852–1853 (O'Connell, P. F., Ed.). Princeton University Press, Princeton, NJ.
Thoreau, H. D. (2013). Walden (Li, J., Trans.). Tianjin People's Publishing House, Tianjin.
Tobia, K. P. (2015). Wonder and value. Res Philosophica, 92(4), 959–984.
Tuan, Y.-F. (1977). Space and place: The perspective of experience. University of Minnesota Press, Minneapolis.
Urbas, Joseph (2021). The Philosophy of Ralph Waldo Emerson. Routledge.
Vallins, D. (2000). Coleridge and the psychology of Romanticism. Palgrave Macmillan, London.
Versluis, Arthur (2014). American Gurus: From Transcendentalism to New Age Religion. Oxford University Press.
Vitz, R. (2015). Reforming the art of living: Nature, virtue, and religion in Descartes's epistemology. Springer, Cham.
Wang, Bo (2018). "Re-discussion on 'Ran' and 'Nature'." Philosophical Research.
Wang, Robin (2012). Yinyang: The Way of Heaven and Earth in Chinese Thought and Culture. Cambridge University Press.
Wang, Q. (2014). Research on Plotinus' eschatological thought. People's Publishing House, Beijing.

Wang, Y., & Xiao, Y. (2023). Chinese Chan Buddhism and the agrarian aesthetic in the garden. Religions. https://www.mdpi.com/journal/religions/special_issues/XOCHHPL3HM. Accessed 28 March 2025.
Wang, Z. (2008). Nature and human beings: The research on the lineage of these two concepts. The Commercial Press, Beijing.
Washington, Haydn (2019). A Sense of Wonder towards Nature: Healing the Planet through Belonging. Routledge.
Wei, Z. (1982). History of Chinese thought (Vol. I). Dalin Publishing House, Taipei.
West, Cornel (1989). The American Evasion of Philosophy: A Genealogy of Pragmatism. University of Wisconsin Press.
West, C. (n.d.). Untitled work (Dong, S., Trans.). Nanjing University Press, Nanjing.
White, L. R., Jr. (1967). The historical roots of our ecological crisis. Science, 155(3767), 1203–1207.
Wolf, S. (2010). Good for nothings: Presidential address delivered at the One Hundred Seventh Annual Meeting of the Eastern Division of the American Philosophical Association. Proceedings and Addresses of the American Philosophical Association, 84(2), 47–64.
Worster, Donald (2008). *A Passion for Nature: The Life of John Muir.* Oxford University Press.
Worster, D. (2019). A passion for nature: The life of John Muir (Trans.). Life·Knowledge·Sanlian Bookstore, Beijing.
Wilderness Act (1964). Public Law 88–577, 88th Congress. Wolf, Susan (2010) Meaning in Life and Why It Matters. Princeton University Press.
Xiaogan Liu, *Interpretation and Orientation: An Inquiry into the Methodology of Research in Chinese Philosophy* (The Commercial Press, 2009). 刘笑敢:《诠释与定向:中国哲学研究方法之探究》,商务印书馆, 2009.
Xie, Z. (2012). Research on American transcendentalism's acceptance of Confucianism. Peking University Press, Beijing.
Xu, S., & Xu, X. (2013). Shuowen Jiezi. Zhonghua Book Company, Beijing.
Yan-Shi-an. (2010). The awareness of evil human nature in Xun Zi, Han Fei Zi, and Zhuangzi: A preliminary analysis. Journal of Nanjing University, 47(2), 63–78.
Yang, R., & Cao, Y. (2008, September 22). Promoting China's wilderness conservation with national parks as an opportunity. Guangming Daily, p. 8.
Yifu Duan. (2001). Space and Place: The Perspective of Experience, University of Minnesota Press.
Yao, D. (1993). Unity of Heaven and man: Spinoza and Confucianism. Journal of Social Sciences of Jilin University, 1993(1), 56–67.
Yu, Ying-shi (2014). *Between Heaven and Man: An Inquiry into the Origins of Ancient Chinese Thought* (论天人之际:中国古代思想起源试探). Zhonghua Book Company.
Yuedi, Liu. Living Aesthetics (2024). Critical Modernity and Reconstruction of Aesthetics, Beijing Normal University Press.
Zhao, D., & Fu, L. (Eds.). (2016). Medieval philosophy. The Commercial Press, Beijing.
Zhaodong, H. (2003). A summary of contemporary Western Spinoza philosophy research. Philosophical Trends, 2003(7), 34–42.
Zhuangzi (1968). *The Complete Works of Zhuangzi,* trans. Burton Watson. Columbia University Press.
Zhuangzi. (2009). *Zhuangzi: The Essential Writings with Selections from Traditional Commentaries,* trans. Brook Ziporyn. Hackett Publishing.

MIX
Papier aus verantwortungsvollen Quellen
Paper from responsible sources
FSC® C105338

If you have any concerns about our products,
you can contact us on
ProductSafety@springernature.com

In case Publisher is established outside the EU,
the EU authorized representative is:
Springer Nature Customer Service Center GmbH
Europaplatz 3, 69115 Heidelberg, Germany

Printed by Libri Plureos GmbH
in Hamburg, Germany